THE PERFECT BET

THE PERFECT

How Science and Math Are Taking the Luck Out of Gambling

Adam Kucharski

BASIC BOOKS
A MEMBER OF THE PERSEUS BOOKS GROUP
NEW YORK

Published by Basic Books,
A Member of the Perseus Books Group

Books published by Basic Books are available at special discounts for bulk purchases in the United States by corporations, institutions, and other organizations. For more information, please contact the Special Markets Department at the Perseus Books Group, 2300 Chestnut Street, Suite 200, Philadelphia, PA 19103, or call (800) 810-4145, ext. 5000, or e-mail special.markets@perseusbooks.com.

Designed by Linda Mark

Library of Congress Cataloging-in-Publication Data
Kucharski, Adam (Mathematician)
The perfect bet : how science and math are taking the luck out of gambling / Adam Kucharski.
pages cm
Includes bibliographical references and index.
ISBN 978-0-465-05595-1 (hardcover)—ISBN 978-0-465-09859-0 (ebook)
1. Games of chance (Mathematics) 2. Gambling. 3. Gambling systems. 4. Probabilities. I. Title.
QA271.K83 2015
519.2'7—dc23

2015034255

10 9 8 7 6 5 4 3 2 1

For my parents

Luck is probability taken personally.

—Chip Denman

CONTENTS

INTRODUCTION

IN JUNE 2009, A BRITISH NEWSPAPER TOLD THE STORY OF ELLIOTT Short, a former financial trader who'd made over £20 million betting on horse races. He had a chauffeur-driven Mercedes, kept an office in the exclusive Knightsbridge district of London, and regularly ran up huge bar tabs in the city's best clubs. According to the article, Short's winning strategy was simple: always bet against the favorite. Because the highest-rated horse doesn't always win, it was possible to make a fortune using this approach. Thanks to his system, Short had made huge profits on some of Britain's best-known races, from £1.5 million at Cheltenham Festival to £3 million at Royal Ascot.

There was just one problem: the story wasn't entirely true. The profitable bets that Short claimed to have made at Cheltenham and Ascot had never been placed. Having persuaded investors to pour hundreds of thousands of pounds into his betting system, he'd spent much of the money on holidays and nights out. Eventually, his investors started asking questions, and Short was arrested. When the case went to trial in April 2013, Short was found guilty of nine counts of fraud and was sentenced to five years in prison.

It might seem surprising that so many people were taken in. But there is something seductive about the idea of a perfect betting system. Stories of successful gambling go against the notion that casinos and bookmakers are unbeatable. They imply that there are flaws in games of chance, and that these can be exploited by anyone sharp enough to spot them. Randomness can be reasoned with, and fortune controlled by formulae. The idea is so appealing that, for as long as many games have existed, people have tried to find ways to beat them. Yet the search for the perfect bet has not only influenced gamblers. Throughout history, wagers have transformed our entire understanding of luck.

WHEN THE FIRST ROULETTE wheels appeared in Parisian casinos in the eighteenth century, it did not take long for players to conjure up new betting systems. Most of the strategies came with attractive names, and atrocious success rates. One was called "the martingale." The system had evolved from a tactic used in bar games and was rumored to be foolproof. As its reputation spread, it became incredibly popular among local players.

The martingale involved placing bets on black or red. The color didn't matter; it was the stake that was important. Rather than betting the same amount each time, a player would double up after a loss. When players eventually picked the right color, they would therefore win back all the money lost on earlier bets plus a profit equal to their initial stake.

At first glance, the system seemed flawless. But it had one major drawback: sometimes the required bet size would increase far beyond what the gambler, or even casino, could afford. Following the martingale might earn a player a small profit initially, but in the long run solvency would always get in the way of strategy. Although the martingale might have been popular, it was a tactic that no one could afford to carry out successfully. "The martingale is as elusive as the soul," as writer Alexandre Dumas put it.

One of the reasons the strategy lured in so many players—and continues to do so—is that mathematically it appears perfect. Write down the amount you've bet and the amount you could win, and you'll always come out on top. The calculations have a flaw only when they meet reality. On paper, the martingale seems to work fine; in practical terms, it's hopeless.

When it comes to gambling, understanding the theory behind a game can make all the difference. But what if that theory hasn't been invented yet? During the Renaissance, Gerolamo Cardano was an avid gambler. Having frittered away his inheritance, he decided to make his fortune by betting. For Cardano, this meant measuring how likely random events were.

Probability as we know it did not exist in Cardano's era. There were no laws about chance events, no rules about how likely something was. If someone rolled two sixes while playing dice, it was simply good luck. For many games, nobody knew precisely what a "fair" wager should be.

Cardano was one of the first to spot that such games could be analyzed mathematically. He realized that navigating the world of chance meant understanding where its boundaries lay. He would therefore look at the collection of all possible outcomes, and then home in on the ones that were of interest. Although two dice could land in thirty-six different arrangements, there was only one way to get two sixes. He also worked out how to deal with multiple random events, deriving "Cardano's formula" to calculate the correct odds for repeated games.

Cardano's intellect was not his only weapon in card games. He also carried a long knife, known as a poniard, and was not opposed to using it. In 1525, he was playing cards in Venice and realized his opponent was cheating. "When I observed that the cards were marked, I impetuously slashed his face with my poniard," Cardano said, "though not deeply."

In the decades that followed, other researchers chipped away at the mysteries of probability, too. At the request of a group of Italian nobles, Galileo investigated why some combinations of dice faces appeared more often than others. Astronomer Johannes Kepler also took time off from studying planetary motion to write a short piece on the theory of dice and gambling.

The science of chance blossomed in 1654 as the result of a gambling question posed by a French writer named Antoine Gombaud. He had been puzzled by the following dice problem. Which is more likely: throwing a single six in four rolls of a single die, or throwing double sixes in twenty-four rolls of two dice? Gombaud believed the two events would occur equally often but could not prove it. He wrote to his mathematician friend Blaise Pascal, asking if this was indeed the case.

To tackle the dice problem, Pascal enlisted the help of Pierre de Fermat, a wealthy lawyer and fellow mathematician. Together, they built on Cardano's earlier work on randomness, gradually pinning down the basic laws of probability. Many of the new concepts would become central to mathematical theory. Among other things, Pascal and Fermat defined the "expected value" of a game, which measured how profitable it would be on average if played repeatedly. Their research showed that Gombaud had been wrong: he was more likely to get a six in four rolls of one die than double sixes in twenty-four rolls of two dice. Still, thanks to Gombaud's gambling puzzle, mathematics had gained an entirely new set of ideas. According to mathematician Richard Epstein, "Gamblers can rightly claim to be the godfathers of probability theory."

As well as helping researchers understand how much a bet is worth in purely mathematical terms, wagers have also revealed how we value decisions in real life. During the eighteenth century, Daniel Bernoulli wondered why people would often prefer low-risk bets to ones that were, in theory, more profitable. If expected profit was not driving their financial choices, what was?

Bernoulli solved the wager problem by thinking in terms of "expected utility" rather than expected payoff. He suggested that the same amount of money is worth more—or less—depending on how much a person already has. For example, a single coin is more valuable to a poor person than it is to a rich one. As fellow researcher Gabriel Cramer said, "The mathematicians estimate money in proportion to its quantity, and men of good sense in proportion to the usage that they may make of it."

Such insights have proved to be very powerful. Indeed, the concept of utility underpins the entire insurance industry. Most people prefer to make regular, predictable payments than to pay nothing and risk getting hit with a massive bill, even if it means paying more on average. Whether we buy an insurance policy or not depends on its utility. If something is relatively cheap to replace, we are less likely to insure it.

Over the following chapters, we will find out how gambling has continued to influence scientific thinking, from game theory and statistics to chaos theory and artificial intelligence. Perhaps it shouldn't be surprising that science and gambling are so intertwined. After all, wagers are windows into the world of chance. They show us how to balance risk against reward and why we value things differently as our circumstances change. They help us to unravel how we make decisions and what we can do to control the influence of luck. Encompassing mathematics, psychology, economics, and physics, gambling is a natural focus for researchers interested in random—or seemingly random—events.

The relationship between science and betting is not only benefiting researchers. Gamblers are increasingly using scientific ideas to develop successful betting strategies. In many cases, the concepts are traveling full circle: methods that originally emerged from academic curiosity about wagers are now feeding back into real-life attempts to beat the house.

THE FIRST TIME PHYSICIST Richard Feynman visited Las Vegas in the late 1940s, he went from game to game, working out how much he could expect to win (or, more likely, lose). He decided that although craps was a bad deal, it wasn't that bad: for every dollar he bet, he could expect to lose 1.4 cents on average. Of course, that was the expected loss over a large number of attempts. When Feynman tried the game, he was particularly unlucky, losing five dollars right away. It was enough to put him off casino gambling for good.

Nevertheless, Feynman made several trips to Vegas over the years. He was particularly fond of chatting with the showgirls. During one trip, he had lunch with a performer named Marilyn. As they were eating, she pointed out a man strolling across the grass. He was a well-known professional gambler named Nick Dandolos, or "Nick the Greek." Feynman found the notion puzzling. Having calculated the odds for each casino game, he couldn't work out how Nick the Greek could consistently make money.

Marilyn called Nick the Greek over to their table, and Feynman asked how it was possible to make a living gambling. "I only bet when the odds are in my favor," Nick replied. Feynman didn't understand what he meant. How could the odds ever be in someone's favor?

Nick the Greek told Feynman the real secret behind his success. "I don't bet on the table," he said. "Instead, I bet with people around the table who have prejudices—superstitious ideas about lucky numbers." Nick knew the casino had the edge, so he made wagers with naive fellow gamblers instead. Unlike the Parisian gamblers who used the martingale strategy, he understood the games, and understood the people playing them. He had looked beyond the obvious strategies—which would lose him money—and found a way to tip the odds in his favor. Working out the numbers hadn't been the tricky part; the real skill was turning that knowledge into an effective strategy.

Although brilliance is generally less common than bravado, stories of other successful gambling strategies have emerged over the years. There are tales of syndicates that have successfully exploited lottery loopholes and teams that have profited from flawed roulette tables. Then there are the students—often of the mathematical variety—who have made small fortunes by counting cards.

Yet in recent years these techniques have been surpassed by more sophisticated ideas. From the statisticians forecasting sports scores to the inventors of the intelligent algorithms that beat human poker players, people are finding new ways to take on casinos and bookmakers. But who are the people turning hard science into hard cash? And—perhaps more importantly—where did their strategies come from?

Coverage of winning exploits often focuses on who the gamblers were or how much they won. Scientific betting methods are presented as mathematical magic tricks. The critical ideas are left unreported; the theories remain buried. But we should be interested in how these tricks are done. Wagers have a long history of inspiring new areas of science and generating insights into luck and decision making. The methods have also permeated wider society, from technology to finance. If we can uncover the inner workings of modern betting strategies, we can find out how scientific approaches are continuing to challenge our notions of chance.

From the simple to the intricate, from the audacious to the absurd, gambling is a production line for surprising ideas. Around the globe, gamblers are dealing with the limits of predictability and the boundary between order and chaos. Some are examining the subtleties of decision making and competition; others are looking at quirks of human behavior and exploring the nature of intelligence. By dissecting successful betting strategies, we can find out how gambling is still influencing our understanding of luck—and how that luck can be tamed.

1

THE THREE DEGREES OF IGNORANCE

BENEATH LONDON'S RITZ HOTEL LIES A HIGH-STAKES CASINO. It's called the Ritz Club, and it prides itself on luxury. Croupiers dressed in black oversee the ornate tables. Renaissance paintings line the walls. Scattered lamps illuminate the gold-trimmed decor. Unfortunately for the casual gambler, the Ritz Club also prides itself on exclusivity. To bet inside, you need to have a membership or a hotel key. And, of course, a healthy bankroll.

One evening in March 2004, a blonde woman walked into the Ritz Club, chaperoned by two men in elegant suits. They were there to play roulette. The group weren't like the other high rollers; they turned down many of the free perks usually doled out to big-money players. Still, their focus paid off, and over the course of the night, they won £100,000. It wasn't exactly a small sum, but it was by no means unusual by Ritz standards. The following night the group returned to the casino and again perched beside a roulette table. This time their winnings were much larger. When they eventually cashed in their chips, they took away £1.2 million.

Casino staff became suspicious. After the gamblers left, security looked at the closed-circuit television footage. What they saw was enough to make them contact the police, and the trio were soon arrested at a hotel not far from the Ritz. The woman, who turned out to be from Hungary, and her accomplices, a pair of Serbians, were accused of obtaining money by deception. According to early media reports, they had used a laser scanner to analyze the roulette table. The measurements were fed into a tiny hidden computer, which converted them into predictions about where the ball would finally land. With a cocktail of gadgetry and glamour, it certainly made for a good story. But a crucial detail was missing from all the accounts. Nobody had explained precisely how it was possible to record the motion of a roulette ball and convert it into a successful prediction. After all, isn't roulette supposed to be random?

THERE ARE TWO WAYS to deal with randomness in roulette, and Henri Poincaré was interested in both of them. It was one of his many interests: in the early twentieth century, pretty much anything that involved mathematics had at some point benefited from Poincaré's attention. He was the last true "Universalist"; no mathematician since has been able to skip through every part of the field, spotting crucial connections along the way, like he did.

As Poincaré saw it, events like roulette appear random because we are ignorant of what causes them. He suggested we could classify problems according to our level of ignorance. If we know an object's exact initial state—such as its position and speed—and what physical laws it follows, we have a textbook physics problem to solve. Poincaré called this the first degree of ignorance: we have all the necessary information; we just need to do a few simple calculations.

The second degree of ignorance is when we know the physical laws but don't know the exact initial state of the object, or cannot measure it accurately. In this case we must either improve our measurements or limit our predictions to what will happen to the object in the very near future. Finally, we have the third, and most extensive, degree of ignorance. This is when we don't know the initial state of the object or the physical laws. We can also fall into the third level of ignorance if the laws are too intricate to fully unravel. For example, suppose we drop a can of paint into a swimming pool. It might be easy to predict the reaction of the swimmers, but predicting the behavior of the individual paint and water molecules will be far more difficult.

We could take another approach, however. We could try to understand the effect of the molecules bouncing into each other without studying the minutiae of the interactions between them. If we look at all the particles together, we will be able to see them mix together until—after a certain period of time—the paint spreads evenly throughout pool. Without knowing anything about the cause, which is too complex to grasp, we can still comment on the eventual effect.

The same can be said for roulette. The trajectory of the ball depends on a number of factors, which we might not be able to grasp simply by glancing at a spinning roulette wheel. Much like for the individual water molecules, we cannot make predictions about a single spin if we do not understand the complex causes behind the ball's trajectory. But, as Poincaré suggested, we don't necessarily have to know what causes the ball to land where it does. Instead, we can simply watch a large number of spins and see what happens.

That is exactly what Albert Hibbs and Roy Walford did in 1947. Hibbs was studying for a math degree at the time, and his friend Walford was a medical student. Taking time off from their studies at the University of Chicago, the pair went to Reno to see whether roulette tables were really as random as casinos thought.

Most roulette tables have kept with the original French design of thirty-eight pockets, with numbers 1 to 36, alternately colored black and red, plus 0 and 00, colored green. The zeros tip the game in the casinos' favor. If we placed a series of one-dollar bets on our favorite number, we could expect to win on average once in every thirty-eight attempts, in which case the casino would pay thirty-six dollars. Over the course of thirty-eight spins, we would therefore put down thirty-eight dollars but would only make thirty-six dollars on average. That translates into a loss of two dollars, or about five cents per spin, over the thirty-eight spins.

The house edge relies on there being an equal chance of the roulette wheel producing each number. But, like any machine, a roulette table can have imperfections or can gradually wear down with use. Hibbs and Walford were on the hunt for such tables, which might not have produced an even distribution of numbers. If one number came up more often than the others, it could work to their advantage. They watched spin after spin, hoping to spot something odd. Which raises the question: What do we actually mean by "odd"?

WHILE POINCARÉ WAS IN France thinking about the origins of randomness, on the other side of the English Channel Karl Pearson was spending his summer holiday flipping coins. By the time the vacation was over, the mathematician had flipped a shilling twenty-five thousand times, diligently recording the results of each throw. Most of the work was done outside, which Pearson said "gave me, I have little doubt, a bad reputation in the neighbourhood where I was staying." As well as experimenting with shillings, Pearson got a colleague to flip a penny more than eight thousand times and repeatedly pull raffle tickets from a bag.

To understand randomness, Pearson believed it was important to collect as much data as possible. As he put it, we have "no absolute knowledge of natural phenomena," just "knowledge of our sensations." And Pearson didn't stop at coin tosses and raffle draws. In search of more data, he turned his attention to the roulette tables of Monte Carlo.

Like Poincaré, Pearson was something of a polymath. In addition to his interest in chance, he wrote plays and poetry and studied physics and philosophy. English by birth, Pearson had traveled widely. He was particularly keen on German culture: when University of Heidelberg admin staff accidently recorded his name as Karl instead of Carl, he kept the new spelling.

Unfortunately, his planned trip to Monte Carlo did not look promising. He knew it would be near impossible to obtain funding for a "research visit" to the casinos of the French Riviera. But perhaps he didn't need to watch the tables. It turned out that the newspaper *Le Monaco* published a record of roulette outcomes every week. Pearson decided to focus on results from a four-week period during the summer of 1892. First he looked at the proportions of red and black outcomes. If a roulette wheel were spun an infinite number of times—and the zeros were ignored—he would have expected the overall ratio of red to black to approach 50/50.

Out of the sixteen thousand or so spins published by *Le Monaco*, 50.15 percent came up red. To work out whether the difference was down to chance, Pearson calculated the amount the observed spins deviated from 50 percent. Then he compared this with the variation that would be expected if the wheels were random. He found that a 0.15 percent difference wasn't particularly unusual, and it certainly didn't give him a reason to doubt the randomness of the wheels.

Red and black might have come up a similar number of times, but Pearson wanted to test other things, too. Next, he looked at how often the same color came up several times in a row. Gamblers can become obsessed with such runs of luck. Take the night of August

18, 1913, when a roulette ball in one of Monte Carlo's casinos landed on black over a dozen times in a row. Gamblers crowded around the table to see what would happen next. Surely another black couldn't appear? As the table spun, people piled their money onto red. The ball landed on black again. More money went on red. Another black appeared. And another. And another. In total, the ball bounced into a black pocket twenty-six times in a row. If the wheel had been random, each spin would have been completely unrelated to the others. A sequence of blacks wouldn't have made a red more likely. Yet the gamblers that evening believed that it would. This psychological bias has since been known as the "Monte Carlo fallacy."

When Pearson compared the length of runs of different colors with the frequencies that he'd expect if the wheels were random, something looked wrong. Runs of two or three of the same color were scarcer than they should have been. And runs of a single color—say, a black sandwiched between two reds—were far too common. Pearson calculated the probability of observing an outcome at least as extreme as this one, assuming that the roulette wheel was truly random. This probability, which he dubbed the *p* value, was tiny. So small, in fact, that Pearson said that even if he'd been watching the Monte Carlo tables since the start of Earth's history, he would not have expected to see a result that extreme. He believed it was conclusive evidence that roulette was not a game of chance.

The discovery infuriated him. He'd hoped that roulette wheels would be a good source of random data and was angry that his giant casino-shaped laboratory was generating unreliable results. "The man of science may proudly predict the results of tossing halfpence," he said, "but the Monte Carlo roulette confounds his theories and mocks at his laws." With the roulette wheels clearly of little use to his research, Pearson suggested that the casinos be closed down and their assets donated to science. However, it later emerged that

Pearson's odd results weren't really due to faulty wheels. Although *Le Monaco* paid reporters to watch the roulette tables and record the outcomes, the reporters had decided it was easier just to make up the numbers.

Unlike the idle journalists, Hibbs and Walford actually watched the roulette wheels when they visited Reno. They discovered that one in four wheels had a bias of some sort. One wheel was especially skewed, so betting on it caused the pair's initial one-hundred-dollar stake to grow rapidly. Reports of their final profits differ, but whatever they made, it was enough to buy a yacht and sail it around the Caribbean for a year.

There are plenty of stories about gamblers who've succeeded using a similar approach. Many have told the tale of the Victorian engineer Joseph Jagger, who made a fortune exploiting a biased wheel in Monte Carlo, and of the Argentine syndicate that cleaned up in government-owned casinos in the early 1950s. We might think that, thanks to Pearson's test, spotting a vulnerable wheel is fairly straightforward. But finding a biased roulette wheel isn't the same as finding a profitable one.

In 1948, a statistician named Allan Wilson recorded the spins of a roulette wheel for twenty-four hours a day over four weeks. When he used Pearson's test to find out whether each number had the same chance of appearing, it was clear the wheel was biased. Yet it wasn't clear how he should bet. When Wilson published his data, he issued a challenge to his gambling-inclined readers. "On what statistical basis," he asked, "should you decide to play a given roulette number?"

It took thirty-five years for a solution to emerge. Mathematician Stewart Ethier eventually realized that the trick wasn't to test for a nonrandom wheel but to test for one that would be favorable when betting. Even if we were to look at a huge number of spins and find substantial evidence that one of the thirty-eight numbers came up more often than others, it might not be enough to make a profit. The number would have to appear on average at least once every thirty-six spins; otherwise, we would still expect to lose out to the casino.

The most common number in Wilson's roulette data was nineteen, but Ethier's test found no evidence that betting on it would be profitable over time. Although it was clear the wheel wasn't random, there didn't seem to be any favorable numbers. Ethier was aware that his method had probably arrived too late for most gamblers: in the years since Hibbs and Walford had won big in Reno, biased wheels had gradually faded into extinction. But roulette did not remain unbeatable for long.

WHEN WE ARE AT our deepest level of ignorance, with causes that are too complex to understand, the only thing we can do is look at a large number of events together and see whether any patterns emerge. As we've seen, this statistical approach can be successful if a roulette wheel is biased. Without knowing anything about the physics of a roulette spin, we can make predictions about what might come up.

But what if there's no bias or insufficient time to collect lots of data? The trio that won at the Ritz didn't watch loads of spins, hoping to identify a biased table. They looked at the trajectory of the roulette ball as it traveled around the wheel. This meant escaping not just Poincaré's third level of ignorance but his second one as well.

This is no small feat. Even if we pick apart the physical processes that cause a roulette ball to follow the path it does, we cannot necessarily predict where it will land. Unlike paint molecules crashing into water, the causes are not too complex to grasp. Instead, the cause can be too small to spot: a tiny difference in the initial speed of the ball makes a big difference to where it finally settles. Poincaré argued that a difference in the starting state of a roulette ball—one so tiny it escapes our attention—can lead to an effect so large we cannot miss it, and then we say that the effect is down to chance.

The problem, which is known as "sensitive dependence on initial conditions," means that even if we collect detailed measurements about a process—whether a roulette spin or a tropical storm—a small oversight could have dramatic consequences. Seventy years before mathematician Edward Lorenz gave a talk asking "Does the flap of a butterfly's wings in Brazil set off a tornado in Texas?" Poincaré had outlined the "butterfly effect."

Lorenz's work, which grew into chaos theory, focused chiefly on prediction. He was motivated by a desire to make better forecasts about the weather and to find a way to see further into the future. Poincaré was interested in the opposite problem: How long does it take for a process to become random? In fact, does the path of a roulette ball ever become truly random?

Poincaré was inspired by roulette, but he made his breakthrough by studying a much grander set of trajectories. During the nineteenth century, astronomers had sketched out the asteroids that lay scattered along the Zodiac. They'd found that these asteroids were pretty much uniformly distributed across the night sky. And Poincaré wanted to work out why this was the case.

He knew that the asteroids must follow Kepler's laws of motion and that it was impossible to know their initial speed. As Poincaré put it, "The Zodiac may be regarded as an immense roulette board on which the Creator has thrown a very great number of small balls." To understand the pattern of the asteroids, Poincaré therefore decided to compare the total distance a hypothetical object travels with the number of times it rotates around a point.

Imagine you unroll an incredibly long, and incredibly smooth, sheet of wallpaper. Laying the sheet flat, you take a marble and set it rolling along the paper. Then you set another going, followed by several more. Some marbles you set rolling quickly, others slowly. Because the wallpaper is smooth, the quick ones soon roll far into the distance, while the slow ones make their way along the sheet much more gradually.

The marbles roll on and on, and after a while you take a snapshot of their current positions. To mark their locations, you make a little cut in the edge of the paper next to each one. Then you remove the marbles and roll the sheet back up. If you look at the edge of the roll, each cut will be equally likely to appear at any position around the circumference. This happens because the length of the sheet—and hence the distance the marbles can travel—is much longer than the diameter of the roll. A small change in the marbles' overall distance has a big effect on where the cuts appear on the circumference. If you wait long enough, this sensitivity to initial conditions will mean that the locations of the cuts will appear random. Poincaré showed the same thing happens with asteroid orbits. Over time, they will end up evenly spread along the Zodiac.

To Poincaré, the Zodiac and the roulette table were merely two illustrations of the same idea. He suggested that after a large number of turns, a roulette ball's finishing position would also be completely random. He pointed out that certain betting options would tumble into the realm of randomness sooner than others. Because roulette slots are alternately colored red and black, predicting which of the two appears meant calculating exactly where the ball will land. This would become extremely difficult after even a spin or two. Other options, such as predicting which half of the table the ball lands in, were less sensitive to initial conditions. It would therefore take a lot of spins before the result becomes as good as random.

Fortunately for gamblers, a roulette ball does not spin for an extremely long period of time (although there is an oft-repeated myth that mathematician Blaise Pascal invented roulette while trying to build a perpetual motion machine). As a result, gamblers can—in theory—avoid falling into Poincaré's second degree of ignorance by measuring the initial path of the roulette ball. They just need to work out what measurements to take.

THE RITZ WASN'T THE first time a story of roulette-tracking technology emerged. Eight years after Hibbs and Walford had exploited that biased wheel in Reno, Edward Thorp sat in a common room at the University of California, Los Angeles, discussing get-rich-quick schemes with his fellow students. It was a glorious Sunday afternoon, and the group was debating how to beat roulette. When one of the others said that casino wheels were generally flawless, something clicked in Thorp's mind. Thorp had just started a PhD in physics, and it occurred to him that beating a robust, well-maintained wheel wasn't really a question of statistics. It was a physics problem. As Thorp put it, "The orbiting roulette ball suddenly seemed like a planet in its stately, precise and predictable path."

In 1955, Thorp got hold of a half-size roulette table and set to work analyzing the spins with a camera and stopwatch. He soon noticed that his particular wheel had so many flaws that it made prediction hopeless. But he persevered and studied the physics of the problem in any way he could. On one occasion, Thorp failed to come to the door when his in-laws arrived for dinner. They eventually found him inside rolling marbles along the kitchen floor in the midst of an experiment to find out how far each would travel.

After completing his PhD, Thorp headed east to work at the Massachusetts Institute of Technology. There he met Claude Shannon, one of the university's academic giants. Over the previous decade, Shannon had pioneered the field of "information theory," which revolutionized how data are stored and communicated; the work would later help pave the way for space missions, mobile phones, and the Internet.

Thorp told Shannon about the roulette predictions, and the professor suggested they continue the work at his house a few miles outside the city. When Thorp entered Shannon's basement, it became clear quite how much Shannon liked gadgets. The room was an inventor's playground. Shannon must have had a $100,000 worth of motors, pulleys, switches, and gears down there. He even had a pair of huge polystyrene "shoes" that allowed him to take strolls on the water of a nearby lake, much to his neighbors' alarm. Before long, Thorp and Shannon had added a $1,500 industry-standard roulette table to the gadget collection.

MOST ROULETTE WHEELS ARE operated in a way that allows gamblers to collect information on the ball's trajectory before they bet. After setting the center of the roulette wheel spinning counterclockwise, the croupier launches the ball in a clockwise direction, sending it circling around the wheel's upper edge. Once the ball has looped around a few times, the croupier calls "no more bets" or—if casinos like their patter to have a hint of Gallic charm—"*rien ne va plus*." Eventually, the ball hits one of the deflectors scattered around the edge of the wheel and drops into a pocket. Unfortunately for gamblers, the ball's trajectory is what mathematicians call "nonlinear": the input (its speed) is not directly proportional to the output (where it lands). In other words, Thorp and Shannon had ended up back in Poincaré's third level of ignorance.

Rather than trying to dig themselves out by deriving equations for the ball's motion, they instead decided to rely on past observations. They ran experiments to see how long a ball traveling at a certain speed would remain on the track and used this information to make predictions. During a spin, they would time how long it took for a ball to travel once around the table and then compared the time to their previous results to estimate when it would hit a deflector.

The calculations needed to be done at the roulette table, so at the end of 1960, Thorp and Shannon built the world's first wearable computer and took it to Vegas. They tested it only once, as the wires were unreliable, needing frequent repairs. Even so, it seemed like the computer could be a successful tool. Because the system handed gamblers an advantage, Shannon thought casinos might abandon roulette once word of the research got out. Secrecy was therefore of the utmost importance. As Thorp recalled, "He mentioned that social network theorists studying the spread of rumors claimed that two people chosen at random in, say, the United States are usually linked by three or fewer acquaintances, or 'three degrees of separation.'" The idea of "six degrees of separation" would eventually creep into popular culture, thanks to a highly publicized 1967 experiment by sociologist Stanley Milgram. In the study, participants were asked to help a letter get to a target recipient by sending it to whichever of their acquaintances they thought were most likely to know the target. On average, the letter passed through the hands of six people before eventually reaching its destination, and the six degrees phenomenon was born. Yet subsequent research has shown that Shannon's suggestion of three degrees of separation was probably closer to the mark. In 2012, researchers analyzing Facebook connections—which are a fairly good proxy for real-life acquaintances—found that there are an average of 3.74 degrees of separation between any two people. Evidently, Shannon's fears were well founded.

TOWARD THE END OF 1977, the New York Academy of Sciences hosted the first major conference on chaos theory. They invited a diverse mix of researchers, including James Yorke, the mathematician who first coined the term "chaotic" to describe ordered yet unpredictable phenomena like roulette and weather, and Robert May, an ecologist studying population dynamics at Princeton University.

Another attendee was a young physicist from the University of California, Santa Cruz. For his PhD, Robert Shaw was studying the motion of running water. But that wasn't the only project he was working on. Along with some fellow students, he'd also been developing a way to take on the casinos of Nevada. They called themselves the "Eudaemons"—a nod to the ancient Greek philosophical notion of happiness—and the group's attempts to beat the house at roulette have since become part of gambling legend.

The project started in late 1975 when Doyne Farmer and Norman Packard, two graduate students at UC Santa Cruz, bought a refurbished roulette wheel. The pair had spent the previous summer toying with betting systems for a variety of games before eventually settling on roulette. Despite Shannon's warnings, Thorp had made a cryptic reference to roulette being beatable in one of his books; this throwaway comment, tucked away toward the end of the text, was enough to persuade Farmer and Packard that roulette was worth further study. Working at night in the university physics lab, they gradually unraveled the physics of a roulette spin. By taking measurements as the ball circled the wheel, they discovered they would be able to glean enough information to make profitable bets.

One of the Eudaemons, Thomas Bass, later documented the group's exploits in his book *The Eudaemonic Pie*. He described how, after honing their calculations, the group hid a computer inside a shoe and used it to predict the ball's path in a number of casinos. But there was one piece of information Bass didn't include: the equations underpinning the Eudaemons' prediction method.

MOST MATHEMATICIANS WITH AN interest in gambling will have heard the story of the Eudaemons. Some will also have wondered whether such prediction is feasible. When a new paper on roulette appeared in the journal *Chaos* in 2012, however, it revealed that someone had finally put the method to the test.

Michael Small had first come across *The Eudaemonic Pie* while working for a South African investment bank. He wasn't a gambler and didn't like casinos. Still, he was curious about the shoe computer. For his PhD, he'd analyzed systems with nonlinear dynamics, a category that roulette fell very nicely into. Ten years passed, and Small moved to Asia to take a job at Hong Kong Polytechnic University. Along with Chi Kong Tse, a fellow researcher in the engineering department, Small decided that building a roulette computer could be a good project for undergraduates.

It might seem strange that it took so long for researchers to publicly test such a well-known roulette strategy. However, it isn't easy to get access to a roulette wheel. Casino games aren't generally on university procurement lists, so there are limited opportunities to study roulette. Pearson relied on dodgy newspaper reports because he couldn't persuade anyone to fund a trip to Monte Carlo, and without Shannon's patronage, Thorp would have struggled to carry out his roulette experiments.

The mathematical nuts and bolts of roulette have also hindered research into the problem. Not because the math behind roulette is too complex but because it's too simple. Journal editors can be picky about the types of scientific papers they publish, and trying to beat roulette with basic physics isn't a topic they usually go for. There has been the occasional article about roulette, such as the paper Thorp published that described his method. But though Thorp gave enough away to persuade readers—including the Eudaemons—that computer-based prediction could be successful, he omitted the details. The crucial calculations were notably absent.

Once Small and Tse had convinced the university to buy a wheel, they got to work trying to reproduce the Eudaemons' prediction method. They started by dividing the trajectory of the ball

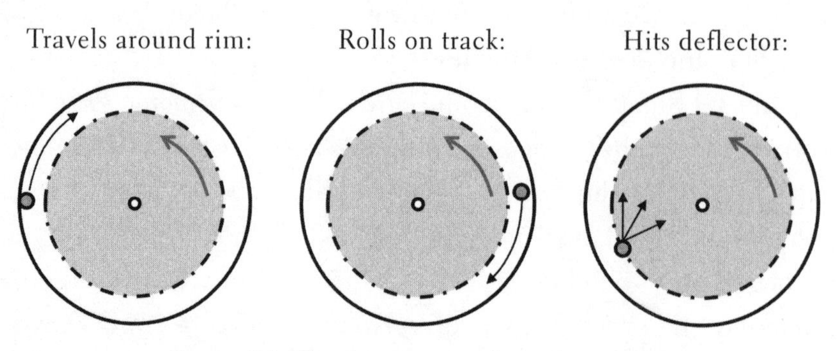

FIGURE 1.1. The three stages of a roulette spin.

into three separate phases. When a croupier sets a roulette wheel in motion, the ball initially rotates around the upper rim while the center of the wheel spins in the opposite direction. During this time, two competing forces act on the ball: centripetal force keeping it on the rim, and gravity pulling it down toward the center of the wheel.

The pair assumed that as the ball rolls, friction slows it down. Eventually, the ball's angular momentum decreases so much that gravity becomes the dominant force. At this point, the ball moves into its second phase. It leaves the rim and rolls freely on the track between the rim and the deflectors. It moves closer to the center of the wheel until it hits one of the deflectors scattered around the circumference.

Until this point, the ball's trajectory can be calculated using textbook physics. But once it hits a deflector, it scatters, potentially landing in one of several pockets. From a betting point of view, the ball leaves a cozy predictable world and moves into a phase that is truly chaotic.

Small and Tse could have used a statistical approach to deal with this uncertainty. However, for the sake of simplicity, they decided to define their prediction as the number the ball was next to when it hit a deflector. To predict the point at which the ball would clip one of the deflectors, Small and Tse needed six pieces of information:

the position, velocity, and acceleration of the ball, and the same for the wheel. Fortunately, these six measurements could be reduced to three if they considered the trajectories from a different standpoint. To an onlooker watching a roulette table, the ball appears to move in one direction and the wheel in the other. But it is also possible to do the calculations from a "ball's-eye view," in which case it's only necessary to measure how the ball moves relative to the wheel. Small and Tse did this by using a stopwatch to clock the times at which the ball passed a specific point.

One afternoon, Small ran an initial series of experiments to test the method. Having written a computer program on his laptop to do the calculations, he set the ball spinning, taking the necessary measurements by hand, as the Eudaemons would have done. As the ball traveled around the rim a dozen or so times, he gathered enough information to make predictions about where it would land. He only had time to run the experiment twenty-two times before he had to leave the office. Out of these attempts, he predicted the correct number three times. Had he just been making random guesses, the probability he would have got at least this many right (the p value) was less than 2 percent. This persuaded him that the Eudaemons' strategy worked. It seemed that roulette really could be beaten with physics.

Having tested the method by hand, Small and Tse set up a high-speed camera to collect more precise measurements about the ball's position. The camera took photos of the wheel at a rate of about ninety frames per second. This made it possible to explore what happened after the ball hit a deflector. With the help of two engineering students, Small and Tse spun the wheel seven hundred times, recording the difference between their prediction and the final outcome. Collecting this information together, they calculated the probability of the ball landing a specified distance away from the predicted pocket. For most of the pockets, this probability wasn't particularly large or small; it was pretty much what they'd have expected if picking pockets at random. Some patterns did emerge, however. The ball landed in the predicted pocket far more often than it would have if the process were down to chance. Moreover, it rarely landed on the numbers that lay on the wheel directly before the predicted pocket. This made sense because the ball would have to bounce backward to get to these pockets.

The camera showed what happened in the ideal situation—when there was very good information about the trajectory of the ball—but most gamblers would struggle to sneak a high-speed camera into a casino. Instead, they would have had to rely on measurements taken by hand. Small and Tse found this wasn't such a disadvantage: they suggested that predictions made with a stopwatch could still provide gamblers with an expected profit of 18 percent.

After announcing his results, Small received messages from gamblers who were using the method in real casinos. "One guy sent me detailed descriptions of his work," he said, "including fabulous photos of a 'clicker' device made from a modified computer mouse strapped to his toe." The work also came to the attention of Doyne Farmer. He was sailing in Florida when heard about Small and Tse's paper. Farmer had kept his method under wraps for over thirty years because—much like Small—he disliked casinos. The trips he made to Nevada during his time with the Eudaemons were enough to convince him that gambling addicts were being exploited by the industry. If people wanted to use computers to beat roulette, he didn't want to say anything that would hand the advantage back to the casinos. However, when Small and Tse's paper was published, Farmer decided it was time to finally break his silence. Especially because there was an important difference between the Eudaemons' approach and the one the Hong Kong researchers had suggested.

Small and Tse had assumed that friction was the main force slowing the ball down, but Farmer disagreed. He'd found that air resistance—not friction—was the main reason for the ball slowing down. Indeed, Farmer pointed out that if we placed a roulette table in a room with no air (and hence no air resistance), the ball would spin around the table thousands of times before settling on a number.

Like Small and Tse's approach, Farmer's method required that certain values be estimated while at the roulette table. During their casino trips, the Eudaemons had three things to pin down: the amount of air resistance, the velocity of the ball when it dropped off the rim of the wheel, and the rate at which the wheel was decelerating. One of the biggest challenges was estimating air resistance and drop velocity. Both influenced the prediction in a similar way: assuming a smaller resistance was much like having an increased velocity.

It was also important to know what was happening around the roulette ball. External factors can have a big effect on a physical process. Take a game of billiards. If you have a perfectly smooth table, a shot will cause the balls to ricochet in a cobweb of collisions. To predict where the cue ball will go after a few seconds, you'd need to know precisely how it was struck. But if you want to make longer-term predictions, Farmer and his colleagues have pointed out it's not enough to merely know about the shot. You also need to take into account forces such as gravity—and not just that of the earth. To predict exactly where the cue ball will travel after one minute, you have to include the gravitational pull of particles at the edge of the galaxy in your calculations.

When making roulette predictions, obtaining correct information about the state of the table is crucial. Even a change in the weather can affect results. The Eudaemons found that if they calibrated their calculations when the weather was sunny in Santa Cruz, the arrival of fog would cause the ball to leave the track half a rotation earlier than they had expected. Other disruptions were closer to home. During one casino visit, Farmer had to abandon betting because an overweight man was resting against the table, tilting the wheel and messing up the predictions.

The biggest hindrance for the group, though, was their technical equipment. They implemented the betting strategy by having one person record the spins and another place the bets, so as not to raise the suspicions of casino security. The idea was that a wireless signal would transmit messages telling the player with the chips which number to bet on. But the system often failed: the signal would disappear, taking the betting instructions with it. Although the group had a 20 percent edge over the casino in theory, these technical problems meant it was never converted into a grand fortune.

As computers have improved, a handful of people have managed to come up with better roulette devices. Most rarely make it into the news, with the exception of the trio who won at the Ritz in 2004. On that occasion, newspapers were particularly quick to latch on to the story of a laser scanner. Yet when journalist Ben Beasley-Murray talked to industry insiders a few months after the incident, they dismissed suggestions that lasers were involved. Instead, it was likely the Ritz gamblers used mobile phones to time the spinning wheel. The basic method would have been similar to the one the Eudaemons used, but advances in technology meant it could be implemented much more effectively. According to ex-Eudaemon Norman Packard, the whole thing would have been pretty easy to set up.

It was also perfectly legal. Although the Ritz group were accused of obtaining money by deception—a form of theft—they hadn't actually tampered with the game. Nobody had interfered with the ball or switched chips. Nine months after the group's initial arrest, police therefore closed the case and returned the £1.3 million haul. In many ways, the trio had the UK's wonderfully archaic gambling laws to thank for their prize. The Gaming Act, which was signed in 1845, had not been updated to cope with the new methods available to gamblers.

Unfortunately, the law does not hand an advantage only to gamblers. The unwritten agreement you have with a casino—pick the correct number and be rewarded with money—is not legally binding in the UK. You can't take a casino to court if you win and it doesn't pay up. And although casinos love gamblers with a losing system, they are less keen on those with winning strategies. Regardless of which strategy you use, you'll have to escape house countermeasures. When Hibbs and Walford passed $5,000 in winnings by hunting for biased tables in Reno, the casino shuffled the roulette tables around to foil them. Even though the Eudaemons didn't need to watch the table for long periods of time, they still had to beat a hasty retreat from casinos on occasion.

AS WELL AS DRAWING the attention of casino security, successful roulette strategies have something else in common: all rely on the fact that casinos believe the wheels are unpredictable. When they aren't, people who have watched the table for long enough can exploit the bias. When the wheel is perfect, and churns out numbers that are uniformly distributed, it can be vulnerable if gamblers collect enough information about the ball's trajectory.

The evolution of successful roulette strategies reflects how the science of chance has developed during the past century. Early efforts to beat roulette involved escaping Poincaré's third level of ignorance, where nothing about the physical process is known. Pearson's work on roulette was purely statistical, aiming to find patterns in data. Later attempts to profit from the game, including the exploits at the Ritz, took a different approach. These strategies tried to overcome Poincaré's second level of ignorance and solve the problem of roulette's outcome being incredibly sensitive to the initial state of the wheel and ball.

For Poincaré, roulette was a way to illustrate his idea that simple physical processes could descend into what seems like randomness. This idea formed a crucial part of chaos theory, which emerged as a new academic field in the 1970s. During this period, roulette was always lurking in the background. In fact, many of the Eudaemons would go on to publish papers on chaotic systems. One of Robert Shaw's projects demonstrated that the steady rhythm of droplets from a dripping tap turns into an unpredictable beat as the tap is unscrewed further. This was one of the first real-life examples of a "chaotic transition" whereby a process switches from a regular pattern to one that is as good as random. Interest in chaos theory and roulette does not appear to have dampened over the years. The topics can still capture the public imagination, as shown by the extensive media attention given to Small and Tse's paper in 2012.

Roulette might be a seductive intellectual challenge, but it isn't the easiest—or most reliable—way to make money. To start with, there is the problem of casino table limits. The Eudaemons played for small stakes, which helped them keep a low profile but also put a cap on potential winnings. Playing at high-stakes tables might bring in more money, but it will also bring additional scrutiny from casino security. Then there are the legal issues. Roulette computers are banned in many countries, and even if they aren't, casinos are understandably hostile toward anyone who uses one. This makes it tricky to earn good profits.

For these reasons, roulette is really only a small part of the scientific betting story. Since the shoe-computer exploits of the Eudaemons, gamblers have been busy tackling other games. Like roulette, many of these games have a long-standing reputation for being unbeatable. And like roulette, people are using scientific approaches to show just how wrong that reputation can be.

2

A BRUTE FORCE BUSINESS

Of the colleges of the University of Cambridge, Gonville and Caius is the fourth oldest, the third richest, and the second biggest producer of Nobel Prize winners. It's also one of the few colleges that serves three-course formal dinners every night, which means that most students end up well acquainted with the college's neo-Gothic dining hall and its unique stained glass windows.

One window depicts a spiraling DNA helix, a nod to former college fellow Francis Crick. Another shows a trio of overlapping circles in tribute to John Venn. There is also a checkerboard situated in the glass, each square colored in a seemingly random way. It's there to commemorate one of the founders of modern statistics, Ronald Fisher.

After winning a scholarship at Gonville and Caius, Fisher spent three years studying at Cambridge, specializing in evolutionary biology. He graduated on the eve of the First World War and tried to join the British Army. Although he completed the medical exams several times, he failed on each occasion because of poor eyesight. As a result, he spent the war teaching mathematics at a number of prominent English private schools, publishing a handful of academic papers in his spare time.

As the conflict drew to a close, Fisher began to search for a new job. One option was to join Karl Pearson's laboratory, where he had been offered the role of chief statistician. Fisher wasn't particularly keen on this option: the previous year, Pearson had published an article criticizing some of his research. Still reeling from the attack, Fisher declined the job.

Instead, Fisher took a job at the Rothamsted Experimental Station, where he turned his attention to agricultural research. Rather than just being interested in the results of experiments, Fisher wanted to make sure that experiments were designed to be as useful as possible. "To consult the statistician after an experiment is finished is often merely to ask him to conduct a post mortem examination," he said. "He can perhaps say what the experiment died of."

Considering the work at hand, Fisher was puzzled about how to scatter different crop treatments across a plot of land during an experiment. The same problem appears when conducting medical trials across a large geographic area. If we are comparing several different treatments, we want to make sure they are scattered across a wide region. But if we distribute them by picking locations at random, there is a chance that we will repeatedly pick similar locations. In which case, a treatment ends up concentrated only in one area, and we have a pretty lousy experiment.

Suppose we want to test four treatments across sixteen trial sites, arranged in a four-by-four grid. How can we scatter the treatments across the area without risking all of them ending up in the same place? In his landmark book *The Design of Experiments*, Fisher suggested that

C	D	B	A
B	A	D	C
D	C	A	B
A	B	C	D

Latin square

A	A	B	D
A	B	D	B
A	C	D	C
D	C	C	B

Lousy square

FIGURE 2.1.

the four treatments be distributed so that they appear in each row and column only once. If the field had good soil at one end and poor land at the other, all treatments therefore would be exposed to both conditions. As it happened, the pattern Fisher proposed had already found popularity elsewhere. It was common in classical architecture, where it was known as a Latin square, as shown in Figure 2.1.

The stained glass window at Gonville and Caius College shows a larger version of a Latin square, with the letters—one for each type of treatment—replaced by colors. As well as earning a tribute in an ancient hall, Fisher's ideas are still used today. The problem of how to construct something that is both random and balanced arises in many industries, including agriculture and medicine. It also comes up in lottery games.

Lotteries are designed to cost players money. They originated as a palatable form of tax, often to support major building projects. The Great Wall of China was financed with profits from a lottery run by the Han dynasty; proceeds from a lottery organized in 1753 funded the British Museum; and many of the Ivy League universities were built on takings from lotteries arranged by colonial governments.

Modern lotteries are made up of several different games, with scratchcards a lucrative part of the business. In the United Kingdom, they make up a quarter of the National Lottery's revenues, and American state lotteries earn tens of billions of dollars from ticket sales. Prizes run into the millions, so lottery operators are careful to limit the supply of winning cards. They can't put random numbers below the scratch-off foil, because there is a chance that could produce more prizes than they could afford to pay out. Nor would it be wise to send batches of cards to places arbitrarily, because one town could end up with all the "lucky" tickets. Scratchcards need to include an element of chance to make sure the game is fair, but operators also need to tweak the game somehow to ensure that there aren't a huge number of winners or too many in one place. To quote statistician William Gossett, they need "controlled randomness."

FOR MOHAN SRIVASTAVA, THE idea that scratchcards follow certain rules started with a joke present. It was June 2003, and he'd been given a handful of cards, including one with a collection of tic-tac-toe games. When he scratched off the foil, he discovered three symbols in a line, which netted him three dollars. It also got him thinking about how the lottery keeps track of the different prizes.

Srivastava worked as a statistician in Toronto, and he suspected that each card contained a code that identified whether it was a winner. Code breaking was something he had always found interesting; he'd known Bill Tutte, the British mathematician who had broken the Nazi Lorenz cipher in 1942, an achievement later described as "one of the greatest intellectual feats of World War II." On the way to collect his prize from a local gas station, Srivastava started to wonder how the lottery might go about distributing the tic-tac-toe scratchcards. He had plenty of experience with such algorithms. He worked as a consultant for mining companies, which hired him to hunt down gold deposits. In high school, he'd even written a computer version of tic-tac-toe for an assignment. He noticed that each foil panel on the scratchcard had a three-by-three grid of numbers printed on it. Perhaps these numbers were the key?

Later that day, Srivastava stopped by the gas station again and bought a bundle of scratchcards. Examining the numbers, he found that some numbers appeared several times on the card, and some only once. As he sifted through a pile of cards, he spotted the fact that if a row contained three of these unique numbers, it usually signaled a winning card. It was a simple and effective method. The challenge was finding such cards.

Unfortunately, it turns out that winning cards aren't all that common. For example, during the early hours of April 16, 2013, a car smashed through the doors of a convenience store in Kentucky. A woman jumped out, grabbed a display containing 1,500 scratchcards, and drove away. By the time she was arrested a few weeks later, she'd managed to claim a mere $200 in prizes.

Even though Srivastava had a reliable—and legal—method for finding profitable scratchcards, it didn't mean he could turn it into a lucrative business. He worked out how long it would take to sort through all potential cards to find the "lucky" ones and realized that he was better off sticking with his existing job. Having decided it wouldn't be worth changing careers, Srivastava thought the lottery might like to know about his discovery. First he tried to get ahold of them by phone, but, perhaps thinking he was just another gambler with a dodgy system, they didn't return his calls. So, he divided twenty untouched scratchcards into two groups—one of winners, and one of losers—and mailed them to the lottery's security team by courier. Srivastava got a phone call from the lottery later that day. "We need to talk," they said.

The tic-tac-toe games were soon removed from stores. According to the lottery, the problem was due to a design flaw. But since

2003, Srivastava has looked at other lotteries in the United States and Canada and suspects some may still be producing scratchcards with the same problem.

In 2011, a few months after *Wired* magazine featured Srivastava's story, reports emerged of an unusually successful scratchcard player in Texas. Joan Ginther had won four jackpots in the Texas scratchcard lottery between 1993 and 2010, bringing in a total of $20.4 million. Was it just down to luck? Although Ginther has never commented on the reason for her multiple wins, some have speculated that her statistics PhD might have had something to do with it.

It's not just scratchcards that are vulnerable to scientific thinking. Traditional lotteries do not include controlled randomness, yet they are still not safe from mathematically inclined players. And when lotteries have a loophole, a winning strategy can begin with something as innocuous as a college project.

EVEN WITHIN A UNIVERSITY as famously offbeat as the Massachusetts Institute of Technology, Random Hall has a reputation for being a little quirky. According to campus legend, the students who first lived there in 1968 wanted to call the dorm "Random House" until the publisher of that name sent them a letter to object. The individual floors have names, too. One is called Destiny, a result of its cash-strapped inhabitants selling the naming rights on eBay; the winning bid was $36 from a man who wanted to name it after his daughter. The hall even has its own student-built website, which allows occupants to check whether the bathrooms or washing machines are available.

In 2005, another plan started to take shape in the corridors of Random Hall. James Harvey was nearing the end of his mathematics degree and needed a project for his final semester. While searching for a topic, he became interested in lotteries.

The Massachusetts State Lottery was set up in 1971 as a way of raising extra revenue for the government. The lottery runs several different games, but the most popular are Powerball and Mega-Millions. Harvey decided that a comparison of the two games could make for a good project. However, the project grew—as projects often do—and Harvey soon began to compare his results with other games, including one called Cash WinFall.

The Massachusetts Lottery introduced Cash WinFall in autumn 2004. Unlike games such as Powerball, which were played in other states too, Cash WinFall was unique to Massachusetts. The rules were simple. Players would choose six numbers for each two-dollar ticket. If they matched all six in the draw, they won a jackpot of at least half a million dollars. If they matched some but not all the numbers, they won a smaller sum. The lottery designed the game so that $1.20 of every $2.00 would be paid out in prizes, with the rest being spent on local good causes. In many ways, WinFall was like all the other lottery games. However, it had one important difference. Usually, when nobody wins the jackpot in a lottery, the prize rolls over to the next draw. If there's no winning ticket next time, it rolls over again and continues to do so until somebody eventually matches all the numbers. The problem with rollovers is that winners—who are good publicity for a lottery—can be rare. And if no smiling faces and giant checks appear in the newspapers for a while, people will stop playing.

Massachusetts Lottery faced precisely that difficulty in 2003, when its Mass Millions game went without a winner for an entire year. They decided that WinFall would avoid this awkward situation by limiting the jackpot. If the prize money rose to $2 million without a winner, the jackpot would "roll down" and instead be split among the players who had matched three, four, or five numbers.

Before each draw, the lottery published its estimate for the jackpot, which was based on ticket sales from previous draws. When the estimated jackpot reached $2 million, players knew that the money would roll down if nobody matched six numbers. People soon spotted that the odds of winning money were far better in a roll-down week than at other times, which meant ticket sales always surged before these draws.

As he studied the game, Harvey realized that it was easier to make money on WinFall than on other lotteries. In fact, the expected payoff was sometimes positive: when a roll down happened, there was at least $2.30 waiting in prize money for every $2.00 ticket sold.

In February 2005, Harvey formed a betting group with some of his fellow MIT students. About fifty people chipped in for the first batch of tickets—raising $1,000 in total—and tripled their money when their numbers came up. Over the next few years, playing the lottery became a full-time job for Harvey. By 2010, he and a fellow team member incorporated the business. They named it Random Strategies Investments, LLC, after their old MIT accommodation.

Other syndicates got in on the action, too. One team consisted of biomedical researchers from Boston University. Another group was led by retired shop owner (and mathematics graduate) Gerald Selbee, who had previously had success with a similar game elsewhere. In 2003, Selbee had noticed a loophole in a Michigan lottery game that also included roll downs. Gathering a thirty-two-person-strong betting group, Selbee spent two years bulk-buying tickets—and netting jackpots—before the lottery was discontinued in 2005. When Selbee's syndicate heard about WinFall, they turned their attention to Massachusetts. There was a good reason for the influx of such betting teams. Cash WinFall had become the most profitable lottery in the United States.

DURING THE SUMMER OF 2010, the WinFall jackpot again approached the roll-down limit. After a $1.59 million prize went unclaimed on August 12, the lottery estimated that the jackpot for the next draw would be around $1.68 million. With a roll down surely only two or three draws away, betting syndicates started to prepare. By the end of the month, they planned to have thousands more dollars in winnings.

But the roll down didn't arrive two draws, or even three draws, later. It came the following week, on August 16. For some reason, there had been a huge increase in ticket sales, enough to drive the total prize money past $2 million. This flood of sales triggered a premature roll down. The lottery officials were as surprised as anyone: they had never sold that many tickets when the estimated jackpot was so low. What was going on?

When WinFall was introduced, the lottery had looked into the possibility of somebody deliberately nudging the draw into a roll down by buying up a large number of tickets. Aware that ticket sales depended on the estimated jackpot—and potential roll downs—the lottery didn't want to get caught out by underestimating the prize money.

They calculated that a player who used stores' automated lottery machines, which churned out tickets with arbitrary numbers, would be able to place one hundred bets per minute. If the jackpot stood at less than $1.7 million, the player would need to buy over five hundred thousand tickets to push it above the $2 million limit. Because this would take well over eighty hours, the lottery didn't think anyone would be able to tip the total over $2 million unless the jackpot was already above $1.7 million.

The MIT group thought otherwise. When James Harvey first started looking at the lottery in 2005, he'd made a trip to the town of Braintree, where the lottery offices were based. He wanted to get ahold of a copy of the guidelines for the game, which would outline precisely how the prize money was distributed. At the time, nobody could help him. But in 2008, the lottery finally sent him the guidelines. The information was a boost for the MIT group, which until then had been relying on their own calculations.

Looking at past draws, the group found that if the jackpot failed to top $1.6 million, the estimate for the next prize was almost always below the crucial value of $2 million. Pushing the draw over the limit on August 16 had been the result of extensive planning. As well as waiting for an appropriate jackpot size—one close to but below $1.6 million—the MIT group had to fill out around 700,000 betting slips, all by hand. "It took us about a year to ramp up to it," Harvey later said. The effort paid off: it's been estimated that they made around $700,000 that week.

Unfortunately, the profits did not continue for much longer. Within a year, the *Boston Globe* had published a story about the loophole in WinFall and the betting syndicates that had profited from it. In the summer of 2011, Gregory Sullivan, Massachusetts Inspector General, compiled a detailed report on the matter. Sullivan pointed out that the actions of the MIT group and others were entirely legal, and he concluded that "no one's odds of having a winning ticket were affected by high-volume betting." Still, it was clear that some people were making a lot of money from WinFall, and the game was gradually phased out.

Even if WinFall hadn't been canceled, the Boston University syndicate told the inspector general that the game wouldn't have remained profitable for betting teams. More people were buying tickets in roll-down weeks, so the prizes were split into smaller and smaller chunks. As the risk of losing money increased, the potential rewards were shrinking. In such a competitive environment, it was crucial to obtain an edge over other teams. The MIT group did this by understanding the game better than many of their competitors: they knew the probabilities and the payoffs and exactly how much advantage they held.

Betting success is not just limited by competition, however. There is also the not-so-small matter of logistics. Gerard Selbee pointed out that if a group wanted to maximize their profits during a roll-down week, they needed to buy 312,000 betting slips, because this was the "statistical sweet spot." The process of buying so many tickets was not always straightforward. The ticket machines would jam in humid weather and run slowly when low on ink. On one occasion, a power outage got in the way of the MIT group's preparations. And some stores would refuse to serve teams altogether.

There was also the question of how to store and organize all the tickets they bought. Syndicates had to keep millions of losing tickets in boxes to show to tax auditors. Moreover, it was a headache to find the winning slips. Selbee claims to have won around $8 million since he starting tackling lotteries in 2003. But after a draw, he and his wife would have to work for ten hours a day examining their collection of tickets to identify the profitable ones.

SYNDICATES HAVE LONG USED the tactic of buying up large combinations of numbers—a method known as a "brute force attack"—to beat lotteries. One of the best-known examples is the story of Stefan Klincewicz, an accountant who hatched a plan to win the Irish National Lottery in 1990. Klincewicz had noticed that it would cost him just under £1 million to buy enough tickets to cover every potential combination, thereby guaranteeing a winning ticket when the draw was made. But the strategy would only work if the jackpot was big enough. While waiting for a large rollover to appear, Klincewicz gathered a twenty-eight-man syndicate. Over a period of six months, the group filled out thousands upon thousands of lottery tickets. When a rollover of £1.7 million was eventually announced for the bank holiday draw in May 1992, they put their plan into action. Picking lottery terminals in quieter locations, the team started placing the necessary bets.

The surge in activity caught the attention of lottery officials, who tried to stop the syndicate by shutting off the terminals they were using. As a result, the group members were able to buy up only 80 percent of the possible number combinations. It wasn't enough to guarantee a win, but it was enough to put luck on their side; when the draw was announced, the syndicate had the winning numbers in their collection. Unfortunately, there were also two other winners, so the group had to share the jackpot. They still ended up with a profit of £310,000, however.

Simple brute force approaches like these do not require many calculations to pull off. The only real obstacle is buying enough tickets. It's more a question of manpower than mathematics, and this reduces the exclusivity of the methods. Whereas roulette players have only to outwit the casino, lottery syndicates often have to compete with other teams attempting to win the same jackpot.

Despite the ongoing competition, some betting syndicates have managed to repeatedly—and legally—turn a profit. Their stories illustrate another difference from roulette betting. Rather than acting alone or in small teams below the official radar, many lottery syndicates have formed companies. They have investors, and they file tax returns. The contrast reflects a wider shift in the world of scientific gambling. What were once individual efforts have grown into an entire industry.

3

FROM LOS ALAMOS TO MONTE CARLO

BILL BENTER IS ONE OF THE WORLD'S MOST SUCCESSFUL GAMBLERS. Based in Hong Kong, his betting syndicate has bet—and won—millions of dollars on horse races over the years. But Benter's gambling career didn't begin with racing. It didn't even begin with sports.

When he was a student, Benter came across a sign in an Atlantic City casino. "Professional card counters are prohibited from playing at our tables." It wasn't a particularly effective deterrent. After reading the sign, only one thought came to mind: card counting works. It was the late 1970s, and casinos had spent the previous decade or so clamping down on a tactic they saw as cheating. Much of the blame—or perhaps credit—for the casinos' losses goes to Edward Thorp. In 1962, Thorp published *Beat the Dealer*, which described a winning strategy for blackjack.

Although Thorp has been called the father of card counting, the idea for a perfect blackjack strategy was actually born in a military barracks. Ten years before Thorp released his book, Private Roger Baldwin had been playing cards with fellow soldiers at the Aberdeen Proving Ground in Maryland. When one of the men suggested a game of blackjack, conversation turned to the rules of the game. They agreed on the basic format. Each player receives two cards, with the dealer getting one faceup and one facedown. Players then choose whether to hit, taking another card in the hope of getting a total bigger than the dealer's, or stand, sticking with their current number. If taking a card sends a player's total over twenty-one, they go "bust" and lose their stake.

Once all players have made their choice, it's the dealer's turn. One of the soldiers pointed out that in Las Vegas, the dealer must stand with a total of seventeen or higher. Baldwin was amazed. The dealer had to follow fixed rules? Whenever he'd played in private games, the dealer was free to do whatever he wanted. Baldwin, who had a master's degree in mathematics, realized this could help him in a casino. If the dealer was subject to strict constraints, it should be possible to find a strategy that would maximize his chances of success.

Like all casino games, blackjack is designed to give the house an edge. Although the dealer and player both appear to have the same aim—drawing cards to get a total near to twenty-one—the dealer has the advantage because the player always goes first. If the player asks for one card too many, and overshoots the target, the dealer wins without doing anything.

Looking at some example blackjack hands, Baldwin noticed his odds improved if he took the value of the dealer's faceup card into account when making decisions. If the dealer had a low card, there was a good chance the dealer would have to draw several cards, increasing the risk of the total going over twenty-one. With a six, for example, the dealer had a 40 percent chance of going bust. With a ten, that probability was halved. Baldwin could therefore get away with standing on a lower total if the dealer had a six, because it was likely that the rules would force the dealer to draw too many cards.

In theory, it would be simple for Baldwin to build these ideas into a perfect strategy. In practice, however, the vast number of potential blackjack hands made the task near impossible to do with pen and paper. To make things worse, a player's choices in a casino weren't just limited to hitting or standing. Players also had the option of doubling their stake, on condition that they would receive one more card to go with the two they already had. Or, if they had received a pair of cards showing the same number, they could "split" these into two separate hands.

Baldwin wouldn't be able to do all the work by hand, so he asked Wilbert Cantey, a sergeant and fellow math graduate, if he could use the base's calculator. Intrigued by Baldwin's idea, Cantey agreed to help, as did James McDermott and Herbert Maisel, two other soldiers who worked in the analytics division.

While Thorp was working on his roulette predictions in Los Angeles, the four men spent their evenings working out the best way to beat the dealer. After several months of calculations, they arrived at what they thought was the optimal strategy. But their perfect system didn't turn out to be, well, perfect. "In statistical terms, we still had a negative expectation," Maisel later said. "Unless you got lucky, you'd still lose in the long run." Even so, by the group's calculations they had managed to reduce the casino's edge to a mere 0.6 percent. In contrast, a player who simply copied the dealer's rules—always standing on seventeen or higher—could expect to lose 6 percent of the time. The four men published their findings in 1956, in a paper titled 'The Optimum Strategy in Blackjack."

It happened that Thorp had already booked a trip to Las Vegas when the paper came out. It was meant to be a relaxing holiday with his wife: a few days of dinner tables rather than blackjack tables. But just before they left, a UCLA professor told Thorp about the soldiers' research. Ever curious, Thorp wrote down the strategy and took it along on his trip.

When Thorp tried the strategy in a casino one evening, slowly reading from a crib sheet while he sat at the table, his fellow gamblers thought he was crazy. Thorp was drawing cards when he should stick and turning down cards when he should take them. He doubled his bet after receiving weak cards. He even chose to split his paltry pair of eights when the dealer had a much stronger hand. What on earth was he thinking?

Despite Thorp's apparently reckless strategy, he didn't run out of chips. One by one, the other players left the table with empty pockets, but Thorp remained. Eventually, having lost eight of his ten dollars, Thorp called it a night. But the little excursion had convinced him that the soldiers' strategy worked better than any other. It also made him wonder how it could be improved.

To simplify calculations, Baldwin had assumed that cards were dealt randomly, with each of the fifty-two cards in the deck having an equal chance of appearing. But blackjack isn't really so random. Unlike roulette, in which each spin is—or at least should be—independent of the last, blackjack has a form of memory: over time, the dealer gradually works through the deck.

Thorp was convinced that if he could record which cards had previously been dealt, it would help him anticipate what might come up next. And because he already had a strategy that in theory pretty much broke even, having information about whether the next card would be high or low was enough to tip the game in his favor. He soon found that even a tactic as simple as keeping track of the number of tens in the deck could turn a profit. By counting cards, Thorp gradually turned the research of the four Aberdeen soldiers—later dubbed the "Four Horsemen of Aberdeen"—into a winning strategy.

Although Thorp made money from blackjack, it wasn't the main reason he made all those trips to Vegas. He saw it more as an academic obligation. When he'd first mentioned the existence of a winning strategy, the response wasn't exactly positive. People ridiculed the idea, just as gamblers had done during his first attempt. After all, Thorp's research challenged the widely held assumption that blackjack couldn't be defeated. *Beat the Dealer* was Thorp's way of proving that his theory was right.

THAT SIGN IN ATLANTIC CITY always stuck with Bill Benter, so when he heard about Thorp's book during a year studying abroad at Bristol University, he headed to the local library to get a copy. He had never seen anything so remarkable. "It showed that nothing was invulnerable," he said. "Old maxims about the house always having the edge were no longer true." Upon his return to the United States, Benter decided to take some time away from study. Switching his university campus in Cleveland, Ohio, for the casinos of Las Vegas, he set to work putting Thorp's system into action. The decision was to prove extremely lucrative: in his early twenties, Benter was making about $80,000 a year from blackjack.

During this time, Benter met an Australian who was also making a tidy sum from card counting. Whereas Benter had gone straight from lecture theaters to casino floors, Alan Woods had started training as an actuary after leaving college. In 1973, his firm was commissioned by the Australian government to calculate the house edge on games in the country's first legal casino. The project introduced Woods to the idea of profitable blackjack systems, and over the next few years he spent his weekends beating casinos around the globe. By the time he met Benter, Woods was a full-time blackjack player. But things were becoming harder for successful gamblers like them.

In the years since Thorp had published his strategy, casinos had become better at spotting card counters. One of the biggest problems with counting—aside from the mental focus required—is that you have to see a lot of cards before you have enough information to make predictions about the rest of the deck. During this time, you have little choice but to use Baldwin's optimal system and bet small amounts to limit your losses. And when you eventually decide that the upcoming cards are likely to be favorable, you need to dramatically increase your stakes to make the most of your advantage. This gives a clear signal to any casino staff on the hunt for card counters. "It's easy to learn how to count cards," as one blackjack professional put it. "It's hard to learn how to get away with it."

Keeping a mental note of card values isn't illegal in Nevada (or anywhere else, for that matter), but that didn't mean Thorp and his strategy were welcome in Las Vegas. Because casinos are private property, they can ban anyone they please. To evade security, Thorp started to wear disguises on his visits. With casinos on the lookout for big changes in betting patterns, gamblers started to search for a better way to play blackjack. Rather than count cards until things looked promising, what if it were possible to predict the order of the entire deck?

MOST MATHEMATICIANS IN THE early twentieth century had read Poincaré's work on probability, but it seemed that hardly anyone truly understood it. One of a few who did was Émile Borel, another mathematician based at the University of Paris. Borel was particularly interested in an analogy Poincaré had used to describe how random interactions—like paint in water—eventually settled down to equilibrium.

Poincaré had compared the situation to the process of card shuffling. If you know the initial order of a deck of cards, randomly swapping a few cards around won't completely mess up the order. Your knowledge about the original deck will therefore still be useful. As more and more shuffles are made, however, this knowledge becomes less and less relevant. Much like paint and water mixing over time, the cards gradually become uniformly distributed, with each card being equally likely to appear at any point in the deck.

Inspired by Poincaré's work, Borel found a way to calculate how quickly the cards would converge to this uniform distribution. His research is still used today when calculating the "mixing time" of a random process, whether card shuffles or chemical interactions. The work also helped blackjack players tackle a growing problem.

To make life difficult for card counters, casinos had started using multiple decks—sometimes combining as many as six—and shuffling the cards before all of them had been dealt. Because this made it harder to keep count, casinos hoped it would help stamp out any player advantage. They didn't realize that the changes also made it harder to shuffle the cards effectively.

During the 1970s, casinos often used a "dovetail shuffle" to mix up the cards. To perform the shuffle, the pack is split in two, and then the two the halves are riffled together. If the cards are perfectly riffled, with cards from each half alternating as they fall, no information is lost: the original order can be recovered by simply looking at every other card. Even if cards fall randomly from each half, however, some information remains.

FIGURE 3.1. A dovetail shuffle. (Credit: Todd Klassy)

Suppose you have a deck of thirteen cards. If you perform a dovetail shuffle, the cards might end up swapping around as follows:

A 2 3 4 5 6 7 8 9 10 J Q K

⇓

A 2 3 4 5 6 7   8 9 10 J Q K

⇓

A 2 **8** 3 **9** **10** 4 5 **J** 6 **Q** **K** 7

The shuffled deck is far from random. Instead, there are two clear sequences of rising numbers (shown in boldface and plain text above). Actually, several card tricks rely on this fact: if a card is placed into an ordered deck and the deck is shuffled once or twice, the extra card will usually stand out because it won't fit into a rising sequence.

For a fifty-two-card deck, mathematicians have shown that a dealer should shuffle the cards at least half a dozen times so as not to leave any detectable patterns. However, Benter found that casinos would rarely bother to be so diligent. Some dealers would shuffle the deck two or three times; others seemed to think that just once was enough.

In the early 1980s, players began to use hidden computers to keep track of the deck. They would enter information by pressing a switch, and the computer would vibrate when a favorable situation cropped up. Keeping track of the shuffles meant it didn't matter whether casinos used several decks. It also helped players avoid giving a clear signal to casino security. If the computer indicated that good cards were likely to come up on the next hand, players didn't have to increase their bets substantially to profit. Unfortunately for gamblers, the advantage no longer exists: computer-aided betting has been illegal in American casinos since 1986.

Even without the clampdown on technology, there was another problem for players like Woods and Benter. Much like Thorp, they had gradually found themselves banned from most casinos around the globe. "Once you become well known," Benter said, "it's a very small world." With casinos refusing to let them play, the pair eventually decided to abandon blackjack. Rather than leave the industry, however, they instead planned to take on a much grander game.

WEDNESDAY NIGHTS AT THE Happy Valley racecourse are busy. Seriously busy. Tucked behind the skyscrapers of Hong Kong Island on a patch of what used to be swampland, over thirty thousand spectators crowd into its stands. Cheers rise above the sound of engines and car horns from the nearby Wan Chai district. The jostling and the noise are signs of how much money is at stake. Gambling is a big part of life at Happy Valley: an average of $145 million was bet during each race day in 2012. To put that in perspective, in that same year the Kentucky Derby set a new American record for betting on a horse race. The total was $133 million.

Happy Valley is managed by the Hong Kong Jockey Club, which also runs the Saturday races at Sha Tin racecourse across the bay in Kowloon. The Jockey Club is a nonprofit organization and has a reputation for running a good operation: gamblers are confident that the races are honest.

Betting in Hong Kong operates on a so-called pari-mutuel system. Rather than gamblers placing a bet with a bookmaker at fixed odds, gamblers' money goes into a pool, with the odds depending on how much has already been wagered on each horse. As an example, suppose there are two horses racing. A total of $200 has been bet on the first, and $300 on the second. Adding these together gives the total betting pool. The race organizers begin by subtracting a fee: in Hong Kong it's 19 percent, which, if the total was $500, would leave

TABLE 3.1. An example tote board.

	Amount bet	Odds
Horse 1	$200	2.03
Horse 2	$300	1.35

$405 in the pot. Then they calculate each horse's odds—the amount you'd get back if you wagered $1 on it—by taking the total available winnings ($405) and dividing it by the amount bet on that horse, as shown in Table 3.1.

Invented by Parisian businessman Joseph Oller, who also founded the Moulin Rouge cabaret club, pari-mutuel betting requires constant calculations and recalculations to produce correct odds. Since 1913, these calculations have been easier thanks to the invention of the "automatic totalizator," commonly known as the "tote board." Its Australian inventor, George Julius, had originally wanted to build a vote-counting machine, but his government had no interest in his design. Undeterred, Julius tweaked the mechanism to calculate betting odds instead and sold it to a racetrack in New Zealand.

In a pari-mutuel system, spectators are effectively betting against each other. The race organizers take the same cut regardless of which horse wins. The odds therefore depend only on which horse the bettors think will do well. Of course, people have all sorts of different methods for picking a successful horse. They might go for one that's been putting in some impressive performances. Perhaps it's won a few races recently or looked confident in practice. It might run well in certain weather. Or have a respected jockey. Maybe it's currently a good weight or a good age.

If enough people bet, we might expect the pari-mutuel odds to settle down to a "fair" value, which reflects the horse's true chances of winning. In other words, the betting market is efficient, bringing together all the scattered bits of information about each horse until there's nothing left to give anyone an advantage. We might expect that to happen, but it doesn't.

When the tote board shows a horse has odds of 100, it suggests bettors think its chance of winning is around 1 percent. Yet it seems people are often too generous about a weaker horse's chances. Statisticians have compared the money people throw at long shots with the amount those horses actually win and have found that the probability of victory is often much lower than the odds imply. Conversely, people tend to underestimate the prospects of the horse that is the favorite to win.

The favorite-long-shot bias means top horses are often more likely to win than their odds suggest. However, betting on them isn't necessarily a good strategy. Because the track takes a cut in a pari-mutuel system, there is a hefty handicap to overcome. Whereas card counters only have to improve on the Four Horsemen's method, which almost breaks even, sports bettors need a strategy that will be profitable even when the track charges 19 percent.

The favorite-long-shot bias might be noticeable, but it's rarely that severe. Nor is it consistent: the bias is larger at some racetracks than at others. Still, it shows that the odds don't always match a horse's chances of winning. Like blackjack, the Happy Valley betting market is vulnerable to smart gamblers. And in the 1980s, it became clear that such vulnerability could be extremely profitable.

HONG KONG WASN'T WOODS'S first attempt at a betting system for horse racing. He'd spent 1982 in New Zealand with a group of professional gamblers, hoping that their collective wisdom would be enough to spot horses with incorrect odds. Unfortunately, it was a year of mixed success.

Benter had a background in physics and an interest in computers, so for the races at Happy Valley, the pair planned to employ a more scientific approach. But winning at the racetrack and winning at blackjack involved very different sets of problems. Could mathematics really help predict horse races?

A visit to the University of Nevada's library brought the answer. In a recent issue of a business journal, Benter spotted an article by Ruth Bolton and Randall Chapman, two researchers based at the University of Alberta in Canada. It was called "Searching for Positive Returns at the Track." In the opening paragraph, they hinted at what followed over the next twenty pages. "If the public makes systematic and detectable errors in establishing the betting odds," they wrote, "it may be possible to exploit such a situation with a superior wagering strategy." Previously published strategies had generally concentrated on well-known discrepancies in racing odds, like the favorite-long-shot bias. Bolton and Chapman had taken a different approach. They'd developed a way to take available information about each horse—such as percentage of races won or average speed—and convert it into an estimate of the probability that horse would win. "It was the paper that launched a multi-billion dollar industry," Benter said. So, how did it work?

TWO YEARS AFTER HIS work on the roulette wheels of Monte Carlo, Karl Pearson met a gentleman by the name of Francis Galton. A cousin of Charles Darwin, Galton shared the family passion for science, adventure, and sideburns. However, Pearson soon noticed a few differences.

When Darwin developed his theory of evolution, he'd taken time to organize the new field, introducing so much structure and direction that his fingerprints can still be seen today. Whereas Darwin was an architect, Galton was an explorer. Much like Poincaré, Galton was happy to announce a new idea and then wander off in search of another. "He never waited to see who was following him," Pearson said. "He pointed out the new land to biologist, to anthropologist, to psychologist, to meteorologist, to economist, and left them to follow or not at their leisure."

Galton also had an interest in statistics. He saw it as a way to understand the biological process of inheritance, a subject that had fascinated him for years. He'd even roped others into studying the topic. In 1875, seven of Galton's friends received sweet pea seeds, with instructions to plant them and return the seeds from their progeny. Some people received heavy seeds; some light ones. Galton wanted to see how the weights of the parent seeds were related to those of the offspring.

Comparing the different sizes of the seeds, Galton found that the offspring were larger than the parents if the parents were small, and smaller than them if the parents were large. Galton called it "regression towards mediocrity." He later noticed the same pattern when he looked at the relationship between heights of human parents and children.

Of course, a child's appearance is the result of several factors. Some of these might be known; others might be hidden. Galton realized it would be impossible to unravel the precise role of each one. But using his new regression analysis, he would be able to see whether some factors contributed more than others. For example, Galton noticed that although parental characteristics were clearly important, sometimes features seemed to skip generations, with characteristics coming from grandparents, or even great-grandparents. Galton believed that each ancestor must contribute some amount to the heritage of a child, so he was delighted when he heard that a horse breeder in Pittsburg, Massachusetts, had published a diagram illustrating the exact process he'd been trying to describe. The breeder, a man by the name of A. J. Meston, used a square to represent the child, and then divided it into smaller squares to show the contribution each ancestor made: the bigger the square, the bigger the contribution. Parents

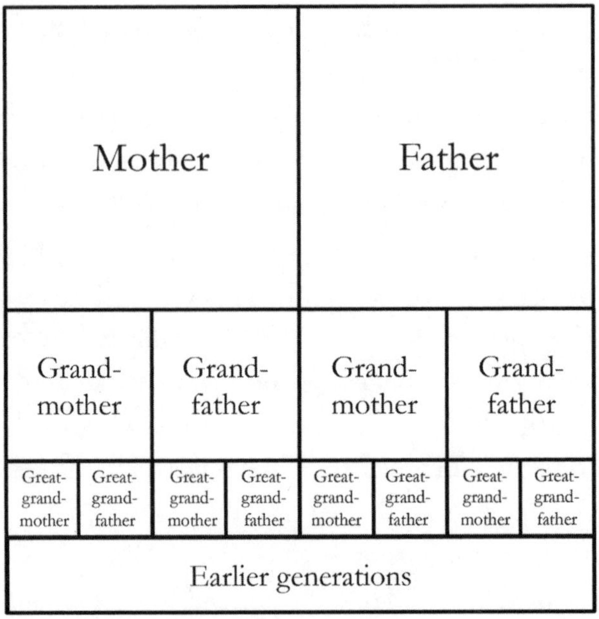

<table>
<tr><td colspan="4">Mother</td><td colspan="4">Father</td></tr>
<tr><td colspan="2">Grand-mother</td><td colspan="2">Grand-father</td><td colspan="2">Grand-mother</td><td colspan="2">Grand-father</td></tr>
<tr><td>Great-grand-mother</td><td>Great-grand-father</td><td>Great-grand-mother</td><td>Great-grand-father</td><td>Great-grand-mother</td><td>Great-grand-father</td><td>Great-grand-mother</td><td>Great-grand-father</td></tr>
<tr><td colspan="8">Earlier generations</td></tr>
</table>

FIGURE 3.2. A. J. Meston's illustration of inheritance.

took up half the space; grandparents a quarter; great-grandparents an eighth, and so on. Galton was so impressed with the idea that he wrote a letter to the journal *Nature* in January 1898 suggesting that they reprint it.

Galton spent a good deal of time thinking about how outcomes, such as child size, were influenced by different factors, and he was meticulous about collecting data to support this research. Unfortunately, his limited mathematical background meant he couldn't take full advantage of the information. When he met Pearson, Galton didn't know how to calculate precisely how much a change in a particular factor would affect the outcome.

Galton had yet again pointed to a new land, and it was Pearson who filled it with mathematical rigor. The pair soon started to apply the ideas to questions about inheritance. Both viewed regression to the mediocre as a potential problem: they wondered how society could make sure that "superior" racial characteristics were not lost in subsequent generations. In Pearson's view, a nation could be improved by "insuring that its numbers are substantially recruited from the better stocks."

From a modern viewpoint, Pearson is a bit of a contradiction. Unlike many of his peers, he thought men and women should be treated as social and intellectual equals. Yet at the same time, he used his statistical methods to argue that certain races were superior to others; he also claimed that laws restricting child labor turned children into social and economic burdens. Today, that's all rather unsavory. Nevertheless, Pearson's work has been hugely influential. Not long after Galton's death in 1911, Pearson established the world's first statistics department at University College London. Building on the diagram Galton had sent to *Nature*, Pearson developed a method for "multiple regression": out of several potentially influential factors, he worked out a way to establish how related each was to a given outcome.

Regression would also provide the backbone for the University of Alberta researchers' racing predictions. Whereas Galton and Pearson used the technique to examine the characteristics of a child, Bolton and Chapman employed it to understand how different factors affected a horse's chances of winning. Was weight more important than percentage of recent races won? How did average speed compare with the reputation of the jockey?

Bolton's first exposure to the world of gambling had come at a young age. "When I was a toddler my Dad took me to the track," she said, "and apparently my little hand picked the winning horse." Despite her early success, it was the last time that she went to the races. Two decades later, however, she found herself picking winners once again, this time with a far more robust method.

The idea for a horseracing prediction method had taken shape in the late 1970s, while Bolton was a student at Queens University in

Canada. Bolton had wanted to learn more about an area of economics known as choice modeling, which aims to capture the benefits and costs of a certain decision. For her final-year dissertation, Bolton teamed up with Chapman, who was researching problems in that area. Chapman, who had a long-standing interest in games, had already accumulated a collection of horse racing data, and together the pair examined how the information could be used to forecast race results. The project was not just the start of an academic partnership; the researchers married in 1981.

Two years after the wedding, Bolton and Chapman submitted the horse racing research to the journal *Management Science*. At the time, prediction methods were growing in popularity, which meant the work received a lot of scrutiny. "The paper spent a long time in review," Bolton said. The research eventually went through four rounds of revisions before appearing in print in the summer of 1986.

In their paper, Bolton and Chapman assumed that a particular horse's chances of winning depended on its quality, which they calculated by bringing together several different measurements. One of these was the starting position. A lower number meant the horse was starting nearer the inside of the track, which should improve a horse's chances, because it means a shorter distance to run. The pair therefore expected regression analysis to show that an increase in starting number would lead to a decrease in quality.

Another factor was the weight of a horse. It was less clear how this would affect quality. Weight restrictions at some races penalize heavier horses, but faster horses often have a higher weight. Old-school racing pundits might try to come up with opinions about which is more important, but Bolton and Chapman didn't need to take such views: they could simply let the regression analysis do the hard work and show them how weight was related to quality.

In Bolton and Chapman's model of a horse race, the quality measurement depended on nine possible factors, including weight, average speed in recent races, and starting position. To illustrate how the different factors contribute to a horse's quality, it's tempting to use a setup similar to the diagram Galton sent *Nature*. However, real life is not as simple as such illustrations suggest. Although Galton's diagram shows how relatives might shape the characteristics of a child, the picture is incomplete because not everything is inherited. Environmental factors can also influence things, and these might not always be visible or known. Moreover, the neat boxes—for mother, father, and so on—are likely to overlap: if a child's father has a certain characteristic, the grandfather or grandmother might have it, too. So, you can't say that each contributing factor is completely independent of the others. The same is true for horse racing. As well as the nine performance-related factors, Bolton and Chapman therefore included an uncertainty factor in their prediction of horse quality. This accounted for unknown influences on horse performance as well as the inevitable quirks of a particular race.

Once the pair had measured the horses' quality, they converted the measurements into predictions about each animal's chance of victory. They did this by calculating the total amount of quality across all the horses in the race. The probability a particular horse would win depended on how much the horse contributed to this overall total.

To work out which factors would be useful for making predictions, Bolton and Chapman compared their model to data from two hundred races. Handling the information was a feat in itself, with race results stored on dozens of computer punch cards. "When I got the data, it was in a big box," Bolton said. "For years, I carried that box around." Getting the results into the computer was also a challenge: it took about an hour to enter the data for each race.

Of the nine factors Bolton and Chapman tested, the pair found that average speed was the most important in deciding where a horse would finish. In contrast, weight didn't seem to make any difference to predictions. Either it was irrelevant or any effect it did have was covered by another factor, in a similar way to how a grandfather's influence on a child's appearance might be covered by the contribution from the father.

It can be surprising which certain factors turn out to be most important. In an early version of Bill Benter's model, the number of races a horse had previously run made a big contribution to the predictions. However, there was no intuitive reason why it was so crucial. Some gamblers might try to think up an explanation, but Benter avoided speculating about specific causes. This is because he knew that different factors were likely to overlap. Rather than try to interpret why something like number of races appears to be important, he instead concentrated on putting together a model that could reproduce the observed race results. Just like the gamblers who searched for biased roulette tables, he could obtain a good prediction without pinning down the precise underlying causes.

In other industries, of course, it might be necessary to isolate how much a certain factor affects an outcome. While Galton and Pearson had been studying inheritance, the Guinness brewery had been trying to improve the life span of its stout. The task fell to William Gossett, a promising young statistician who had spent the winter of 1906 working in Pearson's lab.

Whereas betting syndicates have no control over factors like the weight of a horse, Guinness could alter the ingredients it put in its beer. In 1908, Gossett used regression to see how much hops influenced the drinkable life span of beer. Without hops, the company could expect beer to last between twelve and seventeen days; adding the right amount of hops could increase the life span by several weeks.

Betting teams aren't particularly interested in knowing why certain factors are important, but they do want to know how good their predictions are. It might seem easiest to test the predictions against the racing data the team had just analyzed. Yet this would be an unwise approach.

Before he worked on chaos theory, Edward Lorenz spent the Second World War as a forecaster for the US Air Corps in the Pacific. One autumn in 1944, his team made a series of perfect predictions about weather conditions on the flight path between Siberia and Guam. At least they were perfect according to the reports from aircraft flying that route. Lorenz soon realized what was causing the incredible success rate. The pilots, busy with other tasks, were just repeating the forecast as the observation.

The same problem appears when syndicates test betting predictions against the data used to calibrate the model. In fact, it would be easy to build a seemingly perfect model. For each racing result, they could include a factor that indicates which horse came in first. Then they could tweak these factors until they fitted perfectly with the horses that actually won each race. It would look like they've got a flawless model, when all they've really done is dress up the actual results as a prediction.

If teams want to know how well a strategy will work in the future, they need to see how good it is at predicting *new* events. When collecting information on past races, syndicates therefore put a chunk of the results to one side. They use the rest of the data to evaluate the factors in their model; once this is done, they test the predictions against the collection of yet-to-be-used results. This allows teams to check how the model might perform in real life.

Testing strategies against new data also helps ensure that models satisfy the scientific principle of Occam's razor, which states that if you have to choose between several explanations for an observed event, it is best to pick the simplest. In other words, if you want to build a model of a real-life process, you should shave away the features that you can't justify.

Comparing predictions against new data helps betting teams avoid throwing too many factors into a model, but they still need to assess how good the model actually is. One way to measure the accuracy of a prediction is to use what statisticians call the "coefficient of determination." The coefficient ranges from 0 to 1 and can be thought of as measuring the explanatory power of a model. A value of 0 means that the model doesn't help at all, and bettors might as well pick the winning horse at random; a value of 1 means the predictions line up perfectly with the actual results. Bolton and Chapman's model had a value of 0.09. It was better than randomly choosing horses, but there were still plenty of things that the model wasn't capturing.

Part of the problem was the data they had used. The two hundred races they'd analyzed came from five American racetracks. This meant there was a lot of hidden information: horses would have raced against a range of opponents, in different conditions, with a variety of jockeys. It might have been possible to overcome some of these problems with a lot of racing data, but with only two hundred races? It was doubtful. Still, the strategy could potentially work, if only the race conditions were a bit less variable.

IF YOU HAD TO put together an experiment to study horse racing, it would probably look a lot like Hong Kong. With races happening on one of two tracks, your laboratory conditions are going to be fairly consistent. The subjects of your experiment won't vary too much either: in the United States, tens of thousands of horses race all over the country; in Hong Kong, there is a closed pool of about a thousand horses. With around six hundred races a year, these horses race against each other again and again, which means you can observe similar events several times, just as Pearson always tried to. And, unlike Monte Carlo and its lazy roulette reporters, in Hong Kong there's also plenty of publicly available data on the horses and their performances.

When Benter first analyzed the Hong Kong data, he found that at least five hundred to a thousand races were needed to make good predictions. With fewer than this, there wasn't enough information to work out how much each factor contributed to performance, which meant the model wasn't particularly reliable. In contrast, including more than a thousand races didn't lead to much improvement in the predictions.

In 1994, Benter published a paper outlining his basic betting model. He included a table that showed how his predictions compared to actual race outcomes. The results looked pretty good. Apart from a few discrepancies here and there, the model was remarkably close to reality. However, Benter warned that the results hid a major flaw. If anyone had tried to bet using the predictions, it would have been catastrophic.

SUPPOSE YOU CAME INTO some money and wanted to use the windfall to buy a little bookstore somewhere. There are a couple of ways you could go about it. Having drawn up a short list of stores you might buy, you could go into each one, check the inventory, quiz the management, and examine the accounts. Or you could bypass the paperwork and simply sit outside and count how many customers go in and how many books they come out with. These contrasting strategies reflect the two main ways people approach investing. Researching a company to its core is known as "fundamental analysis," whereas watching how other people view the company over time is referred to as "technical analysis."

Bolton and Chapman's predictions used a fundamental approach. Such methods rely on having good information and sifting through it in the best way possible. The views of pundits don't feature in the analysis. It doesn't matter what other people are doing and which horses they are choosing. The model ignores the betting market. It's like making predictions in a vacuum.

Although it might be possible to make predictions about races in isolation, the same cannot be said for betting on them. If syndicates want to make money at the track, they need to outwit other gamblers. This is where purely fundamental approaches can run into problems. When Benter compared the predictions from his fundamental model with the public odds, he noticed a worrying bias. He'd used the model to find "overlays": horses that, according to the model, have a better chance of winning than their odds imply. These were the horses he would bet on if he were hoping to beat other gamblers. Yet when Benter looked at actual race results, the overlays did not win as often as the predictions suggested. It seemed that the true chances of these horses winning lay somewhere between the probability given by the model and the probability implied by the betting odds. The fundamental approach was clearly missing something.

Even if a betting team has a good model, the public's views on a horse's chances—as shown by the odds on the tote board—aren't completely irrelevant, because not every gambler picks horses based on publicly available information. Some people might know about the jockey's strategy for the race or the horse's eating and workout schedule. When they try to capitalize on this privileged information, it changes the odds on the board.

It makes sense to combine the two available sources of expertise, namely, the model and opinion of other gamblers (as shown by the odds on the tote board). This is the approach Benter advocated. His model still ignores the public odds initially. The first set of predictions is made as if there is no such thing as betting. These predictions are then merged with the public's view. The probability each horse will win is a balance between the chance of the horse winning in the model and the chance of victory according to the current odds. The scales can tip one way or the other: whichever produces the combined prediction that lines up best with actual results. Strike the right balance, and good predictions can become profitable ones.

WHEN WOODS AND BENTER arrived in Hong Kong, they did not meet with immediate success. While Benter spent the first year putting together the statistical model, Woods tried to make money exploiting the long-shot-favorite bias. They had come to Asia with a bankroll of $150,000; within two years, they'd lost it all. It didn't help that investors weren't interested in their strategy. "People had so little faith in the system that they would not have invested for 100 percent of the profits," Woods later said.

By 1986, things were looking better. After writing hundreds of thousands of lines of computer code, Benter's model was ready to go. The team had also collected enough race results to generate decent predictions. Using the model to select horses, they took home $100,000 that year.

Disagreements meant the partnership ended after that first successful season. Before long, Woods and Benter had formed rival syndicates and continued to compete against each other in Hong Kong. Although Woods later admitted that Benter's team had the better model, both groups saw their profits rise dramatically over the next few years.

Several betting syndicates in Hong Kong now use models to predict horse races. Because the track takes a cut, it's difficult to make money on simple bets such as picking the winner. Instead, syndicates chase the more complicated wagers on offer. These include the trifecta: to win, gamblers must predict the horses that will finish first, second, and third in correct order. Then there's the triple trio, which involves winning three trifectas in a row. Although the payoffs for these exotic bets can be huge, the margin for error is also much smaller.

One of the drawbacks with Bolton and Chapman's original model is that it assumes the same level of uncertainty for all the horses. This makes the calculations easier, but it means sacrificing some realism. To illustrate the problem, imagine two horses. The first is a bastion of reliability, always finishing the race in about the same time. The second is more variable, sometimes finishing much quicker than the first, but sometimes taking much longer. As a result, both horses take the same time on average to run a race.

If just these two horses are racing, they will have an equal probability of winning. It might as well be a coin toss. But what if several horses are in the race, each with a different level of uncertainty? If a betting team wants to pick the top three accurately, they need to account for these differences. For years, this was beyond the reach of even the best horse racing models. In the past decade, though, syndicates have found a way to predict races with a varying amount of uncertainty looming over each horse. It's not just recent increases in computing power that have made this possible. The predictions also rely on a much older idea, originally developed by a group of mathematicians working on the hydrogen bomb.

ONE EVENING IN JANUARY 1946, Stanislaw Ulam went to bed with a terrible headache. When he woke up the next morning, he'd lost the ability to speak. He was rushed to a Los Angeles hospital, where concerned surgeons drilled a hole in his skull. Finding his brain severely inflamed as a result of infection, they treated the exposed tissue with penicillin to halt the disease.

Born in Poland, Ulam had left Europe for the United States only weeks before his country fell to the Nazis in September 1939. He was a mathematician by training and had spent most of the Second

World War working on the atomic bomb at Los Alamos National Laboratory. After the conflict ended, Ulam joined UCLA as a professor of mathematics. It wasn't his first choice: amid rumors that Los Alamos might close after the war, Ulam had applied to several higher-profile universities that all turned him down.

By Easter 1946, Ulam had fully recovered from his operation. The stay in the hospital had given him time to consider his options, and he decided to quit his job at UCLA and return to Los Alamos. Far from shutting it down, the government was now pouring money into the laboratory. Much of the effort was going into building a hydrogen bomb, nicknamed the "Super." When Ulam arrived, several obstacles were still in the way. In particular, the researchers needed a means to predict the nuclear chain reactions involved in a detonation. This meant working out how often neutrons collide—and hence how much energy they would give off—inside a bomb. To Ulam's frustration, this couldn't be calculated using conventional mathematics.

Ulam did not enjoy grinding away at problems for hours, as many mathematicians spent their time doing. A colleague once recalled him trying to solve a quadratic equation on a blackboard. "He furrowed his brow in rapt absorption, while scribbling formulas in his tiny handwriting. When he finally got the answer, he turned around and said with relief, 'I feel I have done my work for the day.'"

Ulam preferred to focus on creating new ideas; others could fill in the technical details. It wasn't just mathematical puzzles he tackled in inventive ways. While working at the University of Wisconsin during the winter of 1943, he'd noticed that several of his colleagues were no longer showing up to work. Soon afterward, Ulam received an invitation to join a project in New Mexico. The letter didn't say what was involved. Intrigued, Ulam headed to the campus library and tried to find out all he could about New Mexico. It turned out that there was only one book about the state. Ulam looked at who'd checked it out recently. "Suddenly, I knew where all my friends had disappeared to," he said. Glancing over the others' research interests, he quickly pieced together what they were all working on out in the desert.

WITH HIS HYDROGEN BOMB calculations turning into a series of mathematical cul-de-sacs, Ulam remembered a puzzle he'd thought about during his stay in the hospital. While recovering from surgery, he had passed the time playing solitaire. During one game, he'd tried to work out the probability of a certain card arrangement appearing. Faced with having to calculate a vast set of possibilities—the sort of monotonous work he usually tried to avoid—Ulam realized it might be quicker just to lay out the cards several times and watch what happened. If he repeated the experiment enough times, he would end up with a good idea of the answer without doing a single calculation.

Wondering whether the same technique could also help with the neutron problem, Ulam took the idea to one of his closest colleagues, a mathematician by the name of John von Neumann. The two had known each other for over a decade. It was von Neumann who'd suggested Ulam leave Poland for America in the 1930s; he'd also been the one who invited Ulam to join Los Alamos in 1943. They made quite the pair, portly von Neumann in his immaculate suits—jacket always on—and Ulam with his absent-minded fashion sense and dazzling green eyes.

Von Neumann was quick-witted and logical, sometimes to the point of being blunt. He'd once grown hungry during a train journey and had asked the conductor to send the sandwich seller his way. The request fell on unsympathetic ears. "I will if I see him," the conductor said. To which von Neumann replied, "This train is linear, isn't it?"

When Ulam described his solitaire idea, von Neumann immediately spotted its potential. Enlisting the help of another colleague, a physicist named Nicholas Metropolis, they outlined a way to solve the chain reaction problem by repeatedly simulating neutron collisions. This was possible thanks to the recent construction of a programmable computer at Los Alamos. Because they worked for a government agency, however, the trio needed a code name for their new approach. With a nod to Ulam's heavy-gambling uncle, Metropolis suggested they call it the "Monte Carlo method."

Because the method involved repeated simulations of random events, the group needed access to lots of random numbers. Ulam joked that they should hire people to sit rolling dice all day. His flippancy hinted at an unfortunate truth: generating random numbers was a genuinely difficult task, and they needed a lot of them. Even if those nineteenth-century Monte Carlo journalists had been honest, Karl Pearson would have struggled to build a collection big enough for the Los Alamos men.

Von Neumann, inventive as ever, instead came up with a method for creating "pseudorandom" numbers using simple arithmetic. Despite its being easy to implement, von Neumann knew his method had shortcomings, chiefly the fact that it couldn't generate truly random numbers. "Anyone who considers arithmetical methods of producing random digits is, of course, in a state of sin," he later joked.

As computers have increased in power, and good pseudorandom numbers have become more readily available, the Monte Carlo method has become a valuable tool for scientists. Edward Thorp even used Monte Carlo simulations to produce the strategies in *Beat the Dealer*. However, things aren't so straightforward in horse racing.

In blackjack, only so many combinations of cards can come up—too many to solve the game by hand, but not by computer. Compare this with horse racing models, which can have over a hundred factors. It's possible to tweak the contribution of each one—and hence change the prediction—in a vast number of ways. By just randomly picking different contributions, it's very unlikely you would hit on the best possible model. Every time you made a new guess, it would have the same chance of being the best one, which is hardly the most efficient way of finding the ideal strategy. Ideally, you would make each guess better than the last. This means finding an approach that includes a form of memory.

DURING THE EARLY TWENTIETH century, Poincaré and Borel weren't the only researchers curious about card shuffling. Andrei Markov was a Russian mathematician with a reputation for immense talent and immense temper. When he was young, he'd even picked up the nickname "Andrei Neistovy": Andrei the angry.

In 1907, Markov published a paper about random events that incorporated memory. One example was card shuffling. Just as Thorp would notice decades later, the order of a deck after a shuffle depends on its previous arrangement. Moreover, this memory is short-lived. To predict the effect of the next shuffle, you only need to know the current order; having additional information on the cards' arrangement several shuffles ago is irrelevant. Thanks to Markov's work, this one-step memory has become known as the "Markov property." If the random event is repeated several times, it's a "Markov chain." From card shuffling to Chutes and Ladders, Markov chains are common in games of chance. They can also help when searching for hidden information.

Remember how it takes at least six dovetail shuffles to properly mix up a deck of cards? One of the mathematicians behind that result was a Stanford professor named Persi Diaconis. A few years after Diaconis published his card shuffling paper, a local prison psychologist turned up at Stanford with another mathematical riddle. The psychologist had brought a bundle of coded messages, confiscated from prisoners. Each one was a jumble of symbols made from circles, dots, and lines.

Diaconis decided to give the code to one of his students, Marc Coram, as a challenge. Coram suspected that the messages used a substitution cipher, with each symbol representing a different letter. The difficulty was working out which letter went where. One option was to tackle the problem through trial and error. Coram could have used a computer to shuffle the letters again and again and then examined the resulting text until he hit upon a message that made sense. This is the Monte Carlo method. He would have deciphered the messages eventually, but it could have taken an absurdly long time to get there.

Rather than starting with a new random guess each time, Coram instead chose to use the Markov property of shuffling to gradually improve his guesses. First, he needed a way to measure how realistic a particular guess was. He downloaded a copy of *War and Peace* to find out how often different pairs of letters appeared together. This let him work out how common each particular pairing should be in a given piece of text.

During each round of guessing, Coram randomly switched a couple of the letters in the cipher and checked whether his guess had improved. If a message contained more realistic letter pairings than the previous guess, Coram stuck with it for the next go. If the message wasn't as realistic, he would usually switch back. But occasionally he stuck with a less plausible cipher. It's a bit like solving a Rubik's Cube. Sometimes the quickest route to the solution involves a step that at first glance takes you in the wrong direction. And, like a Rubik's Cube, it might be impossible to find the perfect arrangement by only taking steps that improve things.

The idea of combining the power of the Monte Carlo method with Markov's memory property originated at Los Alamos. When Nick Metropolis first joined the team in 1943, he'd worked on the problem that had also puzzled Poincaré and Borel: how to understand the interactions between individual molecules. It meant solving the equations that described how particles collided, a frustrating task given the crude calculators around at the time.

After years of battling with the problem, Metropolis and his colleagues realized that if they linked the brute force of the Monte Carlo method with a Markov chain, they would be able to infer the properties of substances made of interacting particles. By making smarter guesses, it would be possible to gradually uncover values that couldn't be observed directly. The technique, which became known as "Markov chain Monte Carlo," is the same one Coram would later use to decipher the prison messages.

It eventually took Coram a few thousand rounds of computer-assisted guessing to crack the prison code. This was vastly quicker than a pure brute force method would have been. It turned out that one of the prisoners' messages described the unusual origins of a fight: "Boxer was making loud and loud voices so I tell him por favour can you kick back homie cause I'm playing chess."

To break the prison code, Coram had to take a set of unobserved values (the letters that corresponded to each symbol) and estimate them using letter pairings, which he could observe. In horse racing, betting teams face a similar problem. They don't know how much uncertainty surrounds each horse, or how much each factor should contribute to predictions. But—for a particular level of uncertainty and combination of factors—they can measure how well the resulting predictions match actual race outcomes. The method is classic Ulam. Rather than trying to write down and solve a set of near-impenetrable equations, they let the computer do the work instead.

In recent years, Markov chain Monte Carlo has helped syndicates come up with better race forecasts and predict lucrative exotic results like the triple trio. Yet successful gamblers don't just need to find an edge. They also need to know how to exploit it.

IF YOU WERE BETTING $1.00 on a coin toss coming up tails, a fair payout would be $1.00. Were someone to offer you $2.00 for a bet on tails, that person would be handing you an advantage. You could expect to win $2.00 half the time and suffer a $1.00 loss the other half, which would translate into an expected profit of $0.50.

How much would you bet if someone let you scale up such a biased wager? All of your money? Half of it? Bet too much, and you risk wiping out your savings on an event that still has only a 50 percent chance of success; bet too little, and you won't be fully exploiting your advantage.

After Thorp put together his winning blackjack system, he turned his attention to the problem of such bankroll management. Given a particular edge over the casino, what was the optimal amount to bet? He found the answer in a formula known as the Kelly criterion. The formula is named after John Kelly, a gunslinging Texan physicist who worked with Claude Shannon in the 1950s. Kelly argued that, in the long run, you should wager a percentage of your bankroll equal to your expected profit divided by the amount you'll receive if you win.

For the coin toss above, the Kelly criterion would be the expected payoff ($0.50) divided by the potential winnings ($2.00). This works out to 0.25, which means you should bet a quarter of your available bankroll. In theory, wagering this amount will ensure good profits while limiting the risk of chipping away at your funds. A similar calculation can be performed for horse racing. Betting teams know the probability that a horse will win according to their model. Thanks to the tote board, they can also see what chance the public thinks it has. If the public thinks a victory is less likely than the model suggests, there might be money to be made.

Despite its success in blackjack, there are some drawbacks to the Kelly criterion, especially at the racetrack. First, the calculation assumes you know the true probability of an event. Although you know the chance of a coin coming up heads, things are less clear in horse racing: a model just gives an estimate of the chance a horse will win. If a team overestimates a horse's chances, following the Kelly criterion will cause them to bet too much, increasing their risk of going bust. Consistently overestimate by twofold—for instance, by thinking a horse has a 50 percent chance of victory when in reality it has a 25 percent chance—and it will eventually lead to certain bankruptcy. For this reason, syndicates generally bet less than the Kelly criterion would encourage, often wagering only a half or third of the suggested amount. This reduces the risk of having a "rough ride" and losing a large chunk—or worse, all—of their wealth.

Wagering a smaller amount can also help teams overcome one of the quirks of the betting market in Hong Kong. If you think betting on a certain horse will have a big expected payoff, the Kelly criterion will tell you to put a lot of money on it. In the extreme case, when you are certain of a result, you should stake everything you have. Yet in pari-mutuel betting, this is not necessarily a good idea. A horse's odds depend on the amount wagered, so the more people bet on it, the less money you'll make if it wins.

Even one large bet can shift the whole market. For instance, you might compare your model's predictions and the current odds and notice that you can expect a 20 percent return if you bet on a certain horse. Put down $1.00 and it won't change the overall odds much, so you'll still expect to bag $0.20 if the horse wins. If you have deep pockets, you might choose to bet more than $1.00. The Kelly criterion will certainly be telling you to do so. But if you make a $100.00 bet, it might lower the odds a little. So, your actual profit will be only 19 percent. Still, you've made $19.00.

You might decide to go bigger and bet $1,000. This could shift the odds quite a bit. If a few thousand dollars have already been staked on that horse, it might knock your expected profit down to

10 percent, which means a payoff of $100. Eventually, there comes a point at which putting more money on a horse actually reduces your profits. If the expected return for a $2,000 bet is only 4 percent, you'd be better off wagering a lower amount.

The potential for bets to move the market isn't the only problem you'd have to deal with. All the calculations above assume that you are the last person to bet, and so know the public odds. In reality, devising an optimal strategy isn't that straightforward. At the track, there is a lag on the tote board, sometimes of up to 30 seconds, which means more bets might come in after you've picked your horse.

The total betting pool at Happy Valley might be $300,000 when a team place their bets, but it will probably grow by at least another $100,000 by the time the race starts. Syndicates need to adjust for this influx of cash when deciding how to bet; otherwise, a strategy that initially looks like it will generate a big return could end up producing a mediocre profit. They can't assume that the extra money will be placed on random horses either. In the past decade or so, scientific betting has become more popular, and there are now several syndicates operating in Hong Kong that use models to predict races. These teams are likely to be the ones behind any last-minute betting. "The late money tends to be smart money," Bill Benter said. Teams therefore have to assume the worst: others will also bet on the favorable horse, so any potential profits will have to be divided among more people.

UNTIL SYNDICATES STARTED USING scientific approaches in Hong Kong, successful racetrack betting strategies were few and far between. The techniques are now so effective—and the wins so consistent—that teams like Benter's don't celebrate when their predictions come good. Much of the reason for Benter's early success was the unique setup available to gamblers in Hong Kong. At Happy Valley, gamblers don't have to bet in person at the track. They can call in their selections by phone. This was one of the main reasons Benter and Woods chose Hong Kong. It removed an additional complication and meant they could concentrate on updating their computer predictions rather than worrying about how to place bets in time. Combined with the good availability of data and active betting market, it made Hong Kong the ideal place to implement their strategy.

Gradually, others noticed the appeal of Hong Kong, too. As a result, it is now extremely difficult for betting teams to make money at the city's racetracks. With competition increasing in Hong Kong, the ideas first introduced by Bolton and Chapman are spreading to other regions, including the United States. Over the past decade, scientific betting has become a major part of US horse racing. It's been estimated that teams using computer predictions bet around $2 billion a year at American racetracks, almost 20 percent of the total amount wagered. This sum is all the more impressive when you consider that computer teams cannot bet at several of the large racetracks.

Betting teams are also targeting events in other countries. Like Swedish harness racing, in which horses pull drivers around the track on two-wheeled carts. Imagine a modern version of a Roman chariot race, without the swords and capes. The techniques are growing in popularity at racetracks in Australia and South Africa, too. An idea that began as a piece of academic research has turned into a truly global industry.

It is worth mentioning that it's not cheap to set up a scientific betting syndicate. To gather the necessary technology and expertise—not to mention hone the prediction method and place the bets—costs most teams at least $1 million. Because betting strategies are expensive to run, teams in the United States often seek out racetracks that offer favorable gambling conditions. Several tracks have noticed the bump in profits that comes with the syndicates' huge bets and now encourage computer-based approaches. They even strike deals with betting teams, handing out rebates if the syndicates place large volumes of bets.

These difficulties mean that, although Bolton and Chapman enjoyed the problem-solving aspect of racetrack predictions, they have never really been that interested in gambling careers. Aware of the cost and logistics involved in implementing their strategy, they were happy to remain in academia. "We would joke that we could do it," Bolton said. "Every so often we'd hear how much money was being made and how large these operations had got, but it wasn't for us."

The success of scientific betting in horse racing is all the more remarkable because historically there has been a limit to how much gamblers can predict outcomes. The problem is not limited to horse racing. Whether betting on sports or politics, it has often been difficult to get ahold of the necessary information and to create reliable models. Even if gamblers did manage to come up with a decent prediction, the strategies could be tricky to implement. But at the start of the twenty-first century, that all changed.

4

PUNDITS WITH PHDS

When a new blackjack system hit Britain in 2006, word of its success traveled quietly but quickly. No disguises were required, or card counting, or even visits to casinos. Admittedly, the profit margin was on the sort of scale that would buy pints rather than penthouses, but the system worked. All it required was a computer, a good chunk of spare time, and willingness to do something dull in return for beer money. Students loved it.

The strategy emerged as a result of the new Gambling Act, passed by the government a few months earlier. It meant that UK-based companies could now provide online casino games as well as traditional sports betting. In the rush for new customers, firms started offering signup bonuses. Bet £100 and get an extra £50 free—that sort of thing. At first glance, such a bonus doesn't seem to help much with blackjack. In an online game, it's much easier for casinos to ensure that cards are dealt randomly, making card counting impossible.

If you use the Four Horsemen's optimal blackjack strategy instead, taking the dealer's card into account when making your decision, you can expect to lose money over time. But the signup bonuses tipped things back in the players' favor. People realized that the bonuses would in effect subsidize any losses. Playing the ideal strategy, players would probably lose some of the £100—but not much—and once they'd bet the required total, they would get the bonus. They would usually have to bet this, too, before it could be withdrawn; fortunately, they could simply repeat the previous approach to limit their losses.

During 2006, gamblers hopped from website to website, sitting through hundreds of blackjack hands to build a collection of bonus money. It didn't take long for betting companies to clamp down on what they called "bonus abuse" and exclude games like blackjack from their signup offers. Although there is nothing illegal about setting up a single account to obtain a bonus—indeed, that's the point of a signup bonus—some gamblers pushed the advantage too far. The first conviction for bonus abuse came in the spring of 2012, when Londoner Andrei Osipau was jailed for three years for using fraudulent passports and identity cards to open multiple betting accounts. For those who operated within the law in 2006, profits were far more modest than the £80,000 Osipau was reported to have made. Still, the fact that these bonuses could be exploited illustrates three crucial advantages that gamblers have gained in recent years.

First, the explosion of online betting has meant a far wider range of games and gambling options. In real-life casinos, new games are generally good news for gamblers. According to professional gambler Richard Munchkin, casinos rarely understand how much of an advantage they are offering when they introduce new games. The blackjack loophole that appeared in 2006 showed that the same is true in online gambling. And when the Internet is involved, news of a successful strategy travels much, much faster. The second advantage is the ease with which gamblers can implement a potentially profitable system. Rather than having to dodge casino security or visit bookmakers, they can simply place bets online. Whether through websites or instant messaging, access is quicker and easier than ever before. Finally, the Internet has made it much easier to get ahold of the vital ingredient for many successful betting recipes. From roulette to horse racing, the limited availability of data has dictated where and how people gamble. But these limitations are fading away. As a result, people are targeting a whole host of new games.

EVERY AUTUMN, RECRUITMENT TEAMS descend on the world's best mathematics departments. Most are from the usual crowd: oil firms wanting fluid-dynamics researchers or banks trying to find specialists in probability theory. But in recent years, another type of firm has started to appear at the career events hosted by British universities. Instead of discussing business or finance, they focus on sports such as soccer. Their career presentations are rather like watching a very technical prematch analysis. Formulae and data tables—which most companies hide from prospective applicants—fill the talks. The events have more in common with a lecture than a job pitch.

Many of the approaches are familiar to mathematicians. But although researchers might use the techniques to study ice sheets or epidemics, these firms have found a very different application for the methods. They are using scientific methods to take on the bookmakers. And they are winning.

Modern soccer predictions began with what would otherwise have been a throwaway exam question. During the 1990s, Stuart Coles was a lecturer at Lancaster University, a few miles from the sweeping fells of England's Lake District. Coles specialized in extreme value theory, which deals with the sort of severe, rare events that are like nothing ever seen before. Pioneered by Ronald Fisher in the 1930s, extreme value theory is used to predict the worst-of-the-worst-case scenarios, from floods and earthquakes to wildfires and insurance losses. In short, it is the science of the very unlikely.

Coles's research spanned everything from storm surges to severe pollution. At the prompting of Mark Dixon, another researcher in the department, Coles also started to think about soccer. Dixon had become interested in the topic after looking at a statistics exam given to final-year students at Lancaster. One of the questions involved predicting the results of a hypothetical soccer match, but Dixon spotted a flaw: the method was too simple to be useful in real life. It was an interesting problem, though, and if the ideas were extended—and applied to actual soccer leagues—it might lead to an effective betting strategy.

It took a couple of years for Dixon and Coles to develop the new method and get it ready for publication. The work eventually appeared in the *Journal of Applied Statistics* in 1997. With the research finished, Coles went back to his other projects. Little did he realize how important the soccer paper would turn out to be. "It was one of those things that at the time seemed inconsequential," he said, "but looking back it had a massive impact on my life."

TO PREDICT HORSE RACES in Hong Kong, scientific betting teams assess the quality of each horse and then compare these different quality measurements to work out the probable result. It's tricky to do the same in soccer. Although it might be possible to weigh up each team's qualities and calculate which team is likely to be successful over an entire season, it is much harder to work out who is likely to win in a given match. A team that plays well against one set of opposition can look sluggish against another. Or one shot might go in while another bounces off the woodwork. Then, you have the players. Sometimes a talismanic performance will lift a whole team; sometimes a team will carry along weak players. This tangle of on-pitch activity means that things are much messier from a statistical point of view. During the 1970s, a few researchers had even come to the conclusion that a single soccer match was so dominated by chance that prediction was hopeless.

By choosing to study soccer matches, Dixon and Coles were clearly walking into difficult territory. There was one thing on their side, however. In the United Kingdom, betting odds were generally fixed several days ahead of the match. Unlike the hectic last-minute betting at Hong Kong's racetracks, anyone analyzing soccer matches would have plenty of time to come up with a prediction and compare it to the bookmakers' odds. Even better, there were plenty of potential wagers available. Thanks to a well-established soccer betting market in the United Kingdom, there are all sorts of things to bet on, from half-time score to the number of corner kicks.

Dixon and Coles chose to start with the big question: Which team was going to win? Rather than trying to predict the final result directly, they decided to estimate the number of goals that would be scored before the final whistle. To keep things simple, the pair assumed that each team would score goals at some fixed rate over the course of a game and that the probability of scoring at each point in time was independent of what had already happened in the match.

Events that obey such rules are said to follow a "Poisson process." Named after physicist Siméon Poisson, the process crops up in many walks of life. Researchers have used the Poisson process to model telephone calls to a switchboard, radioactive decay, and even neuron activity. If you assume something follows a Poisson process, you are assuming that events occur at a fixed rate. The world has no memory; each time period is independent of the others. If a match is goalless at half time, it won't make a second-half goal more likely.

Having chosen to model a soccer game as a Poisson process—and hence assuming that goals are scored at a consistent rate over the course of the match—Dixon and Coles still needed to know what the scoring rate should be. The number of goals in a match would probably vary depending on who is playing. How many goals should they expect each team to score?

Early in their 1997 paper, Dixon and Coles set out the things you need to do if you want to build a model of a soccer league. First, you need to somehow measure each team's ability. One option is to use some sort of ranking system. Perhaps you could hand teams a certain number of points after each match and then add up the total points earned over a given time period. Most soccer leagues hand out three points for a win, one for a draw, and nothing for a loss, for example. Representing each team's ability with a single number might show which team is doing well, but it's not always possible to convert rankings into good predictions. A 2009 study by Christoph Leitner and colleagues at Vienna University of Economics and Business provided a good illustration of the problem; they came up with forecasts for the Euro 2008 soccer tournament using rankings published by the sport's governing body Fédération Internationale de Football Association (FIFA) and found that the bookmakers' predictions turned out to be far more accurate. To make money betting on soccer, it seems that you need more than one measurement for each team.

Dixon and Coles suggested splitting ability into two factors: attack and defense. Attacking ability reflected a team's aptitude at scoring goals; defensive weakness indicated how poor they were at stopping them. Given a home team with a certain attacking ability and an away team with a certain defensive weakness, Dixon and Coles assumed that the expected number of goals scored by the home team was the product of three factors:

Home attacking ability × Away defensive weakness × Home advantage factor

Here the "home advantage factor" accounts for the boost teams often get when playing at home. In a similar fashion, the expected number of away goals was equal to the away team's attacking ability multiplied by the home defensive weakness (the away team didn't get any extra advantage).

To estimate each team's attacking and defensive prowess, Dixon and Coles collected several years of data on English soccer games from the top four divisions, which among them contained a total of 92 teams. Because the model included an attack and defensive ability for each team, plus an extra factor that specified the home advantage, this meant estimating a total of 185 factors. If every team had played every other team the same number of times, estimation would have been relatively straightforward. However, promotions and relegations—not to mention cup games—meant some matchups were more common than others. Much like the races at Happy Valley, there was too much hidden information for simple calculations. To estimate each of the 185 factors, it was therefore necessary to enlist help from computational methods like the ones developed by the researchers at Los Alamos.

When Dixon and Coles used their model to make predictions about games that had been played in the 1995–1996 season, they found that the forecasts lined up nicely with the actual results. But would the model have been good enough to bet with? They tested it by going through all the games and applying a simple rule: if the model said a particular result was 10 percent more likely than the bookmakers' odds implied, it was worth betting on. Despite using a basic model and betting strategy, the results suggested that the model would be capable of outperforming the bookmakers.

Not long after publishing their work, Dixon and Coles went their separate ways. Dixon set up Atass Sports, a consultancy firm that specialized in the prediction of sports results. Later, Coles would join Smartodds, a London-based company that also worked on sports models. There are now several firms working on soccer prediction, but Dixon and Coles's research remains at the heart of many models. "Those papers are still the main starting points," said David Hastie, who co-founded soccer analytics firm Onside Analysis.

As with any model, though, the research has some weaknesses. "It's not an entirely polished piece of work," Coles has pointed out. One problem is that the measurements for teams' attacking and defending abilities don't change over the course of a game. In reality, players may tire or launch more attacks at certain points in the game. Another issue is that, in real life, draws are more common than a Poisson process would predict. One explanation might be that teams that are trailing put in more effort, with the hope of leveling the score line, whereas their opponents get complacent. But, according to Andreas Heuer and Oliver Rubner, two researchers at the University of Münster, there's something else going on. They reckon the large number of draws is because teams tend to take fewer risks—and hence are less likely to score—if the score line is even in the later stages of a game. When the pair looked at matches in the German Bundesliga from 1968 to 2011, they found that the goal-scoring rate decreased when the score was a draw. This was especially noticeable when the score was 0–0, with players preferring to settle for the "coziness of a draw."

It turned out that certain points in a game created particularly draw-friendly conditions. Heuer and Rubner found that Bundesliga goals tended to follow a Poisson process during the first eighty minutes of the match, with teams finding the net at a fairly consistent rate. It was only during the last period of play that things became more erratic, especially if the away team was leading by one or two goals in the dying minutes of the match.

By adjusting for these types of quirks, sports prediction firms have built on the work of Dixon, Coles, and others and have turned soccer betting into a profitable business. In recent years, these companies have greatly expanded their operations. But though the industry has grown, and new firms have appeared, the scientific betting industry is still relatively new in the United Kingdom. Even the most established firms started post-2000. In the United States, however, sports prediction has a much richer history—sometimes quite literally.

TO PASS TIME DURING dull high school classes, Michael Kent would often read the sports section of the newspaper. Despite living in Chicago, he followed college athletics from all over the country. As he leafed through the scores, he would get to thinking about the winning margin in each game. "A team would beat another team 28–12," he recalled, "and I would say, well how good is that?"

After high school, Kent completed a degree in mathematics before joining the Westinghouse Corporation. He spent the 1970s working in the corporation's Atomic Power Laboratory in Pittsburgh, Pennsylvania, where they designed nuclear reactors for the US Navy. It was very much a research environment: a mixture of mathematicians, engineers, and computer specialists. Kent spent the next few years trying to simulate what happens to a nuclear reactor that has coolant flowing through its fuel channels. In his spare time, he also started writing computer programs to analyze US football games. In many ways, Kent's model did for college sports what Bill Benter's did for horse races. Kent gathered together lots of factors that might influence a game's result, and then used regression to work out which were important. Just as Benter would later do, Kent waited until he had his own estimate before he looked at the betting market. "You need to make your own number," Kent said. "Then—and only then—do you look at what other people have."

STATISTICS AND DATA HAVE long been an important part of American sport. They are particularly prominent in baseball. One reason is the structure of the game: it is split into lots of short intervals, which, as well as providing plenty of opportunities to grab a hot-dog, makes the game much easier to analyze. Moreover, baseball innings can be broken down into individual battles—such as pitcher versus batter—that are relatively independent, and hence statistician-friendly.

Most of the stats that baseball fans pore over today—from batting averages to runs scored—were devised in the nineteenth century by Henry Chadwick, a sports writer who'd honed his ideas watching cricket matches in England. With the growth of computers in the 1970s, it became easier to collate results, and people gradually formed organizations to encourage research into sports statistics. One such organization was the Society for American Baseball Research, founded in 1971. Because the society's acronym was SABR, the scientific analysis of baseball became known as "sabermetrics."

Sports statistics grew in popularity during the 1970s, but several other ingredients are needed to cook up an effective betting strategy. It just so happened that Michael Kent had all of them. "I was very fortunate that a whole bunch of things came together," he said. The first ingredient was data. Not far from Kent's atomic laboratory in Pittsburgh was the Carnegie Library, which had a set of anthologies containing several years' worth of college sports scores and schedules. The good news was that these provided Kent's model with information it needed to generate robust predictions; the bad news was that each result had to be input manually. Kent also had the technology to power the model, with access to the high-speed computer at Westinghouse. His university had been one of the first in the country to get a computer, so Kent already had far more programming experience than most. That wasn't all. As well as knowing how to write computer programs, Kent understood the statistical theory behind his models. At Westinghouse, he'd worked with an engineer named Carl Friedrich, who'd shown him how to create fast, reliable computer models. "He was one of the most brilliant people I ever met," Kent said. "The guy was unbelievable."

Even with the crucial components in place, Kent's gambling career didn't get off to the best start. "Very early on, I had four huge bets," he said. "I lost them all. I lost $5,000 that Saturday." Still, he realized that the misfortunes did have some benefits. "Nothing motivated me more than losing." After working on his model at night for seven years, Kent finally decided to make sports betting his full-time job in 1979. While Bill Benter was making his first forays into blackjack, Kent left Westinghouse for Las Vegas, ready for the new college football season.

Life in the city involved a lot of new challenges. One of them was the logistics of placing the actual bets. It wasn't like Hong Kong, where bettors could simply phone in their selections. In Las Vegas, gamblers had to turn up at a casino with hard currency. Naturally, this made Kent a little nervous. He came to rely on valet parking, because it stopped him having to walk through poorly lit parking lots with tens of thousands of dollars in cash.

Because it was tricky to place bets, Kent teamed up with Billy Walters, a veteran gambler who knew how Las Vegas worked and how to make it work for them. With Walters taking care of the betting, Kent could focus on the predictions. Over the next few years, other gamblers joined them to help implement the strategy. Some assisted with the computer model, while others dealt with the bookmakers. Together, they were known as the "Computer Group," a name that would become admired by bettors almost as much as it was dreaded by casinos.

Thanks to Kent's scientific approach, the Computer Group's predictions were consistently better than Las Vegas bookmakers'. The success also brought some unwanted attention. Throughout the 1980s, the FBI suspected the group was operating illegally, conducting investigations that were driven partly by bemusement at how the group was making so much money. Despite years of scrutiny, however, the investigations didn't come to anything. There were FBI raids, and several members of the Computer Group were indicted, but all were eventually acquitted.

It has been estimated that between 1980 and 1985, the Computer Group placed over $135 million worth of bets, turning a profit of almost $14 million. There wasn't a single year in which they made a loss. The group eventually disbanded in 1987, but Kent would continue to bet on sports for the next two decades. Kent said the division of labor remained much the same: he would come up with the forecasts, and Walters would implement the betting. Kent pointed out that much of the success of his predictions came from the attention he put into the computer models. "It's the model-building that's important," he said. "You have to know how to build a model. And you never stop building the model."

Kent generally worked alone on his predictions, but he did get help with one sport. An economist at a major university on the West Coast came up with football predictions each week. The man was very private about his betting research, and Kent referred to him only as "Professor number 1." Although the economist's estimates were very good, they were different from Kent's forecasts. So, between 1990 and 2005, they would often merge the two predictions.

Kent made his name—and his money—predicting college sports such as football and basketball. But not all sports have received this level of attention. Whereas Kent was coming up with profitable football models in the 1970s, it wasn't until 1998 that Dixon and Coles sketched out a viable method for soccer betting. And some sports are even harder to predict.

ONE AFTERNOON IN JANUARY 1951, Françoise Ulam came home to find her husband Stanislaw staring out of the window. His expression was peculiar, his eyes unfocused on the garden outside. "I found a way to make it work," he said. Françoise asked him what he meant. "The Super," he replied. "It is a totally different scheme, and it will change the course of history."

Ulam was referring to the hydrogen bomb they had developed at Los Alamos. Thanks to the Monte Carlo method and other technological advances, the United States possessed the most powerful weapon that ever existed. It was the early stages of the Cold War, and Russia had fallen behind in the nuclear arms race.

Yet grand nuclear ideas weren't the only innovations appearing during this period. While Ulam had been working on the Monte Carlo method in 1947, a very different kind of weapon had emerged on the other side of the Iron Curtain. It was called the "Avtomat Kalashnikova" after its designer Mikhail Kalashnikov. In subsequent years, the world would come to know it by another name: the AK-47. Along with the hydrogen bomb, the rifle would shape the course of the Cold War. From Vietnam to Afghanistan, it passed through the hands of soldiers, guerrillas, and revolutionaries. The gun is still in use today, with an estimated 75 million AK-47s having been built to date. The main reason for the weapon's popularity lies in its simplicity. It has only eight moving parts, which means it's reliable and easy to repair. It might not be that accurate, but it rarely jams and can survive decades of use.

When it comes to building machines, the fewer parts there are, the more efficient the machine is. Complexity means more friction between the different components: for example, around 10 percent of a car engine's power is wasted because of such friction. Complexity also leads to malfunctions. During the Cold War, expensive Western rifles would jam while the simple AK-47 continued to function. The same is true of many other processes. Making things more complicated often removes efficiency and increases error. Take blackjack: the more cards a dealer uses, the harder it is to shuffle properly. Complexity also makes it harder to come up with accurate forecasts about the future. The more parts that are involved, and the more interactions going on, the harder it is to predict what will happen from limited past data. And when it comes to sport, there is one activity that involves a particularly large number of interactions, which can make predictions very difficult.

US President Woodrow Wilson once described golf as "an ineffectual attempt to put an elusive ball into an obscure hole with an implement ill adapted to the purpose." As well as having to deal with ballistics, golfers must also contend with their surroundings. Golf courses are littered with obstacles, ranging from trees and ponds to sand bunkers and caddies. As a result, the shadow of luck is never far away. A player might hit a brilliant shot, sending the ball toward the hole, only to see it collide with the flagstick and ricochet into a bunker. Or a player could slice the ball into a tree and have it bounce back into a strong position. Such mishaps are so common in golf that the rulebook even has a phrase to cover them. If the ball hits a random object or otherwise goes astray by accident, it's just the "rub of the green."

Whereas horse races in Hong Kong resemble a well-designed science experiment, golf tournaments are more likely to require one of Ronald Fisher's statistical postmortems. Over the four days of a tournament, players tee off at all sorts of different times. The location of the hole also changes between rounds—and if the tournament is in the United Kingdom, so will the weather. If that isn't bad enough, the field of potential winners is huge in a golf tournament. Whereas the Rugby World Cup has twenty teams competing for the trophy, and the UK Grand National has forty horses running, ninety-five players compete for the US Masters each year, and the other three majors are even larger.

All these factors mean that golf is particularly difficult to predict accurately. Golf has therefore been a bit of an outlier in terms of sports forecasting. Some firms are rising to the challenge—Smartodds now has statisticians working on golf prediction—but in terms of betting activity, the sport still lags far behind many others.

Even among different team sports, some games are easier to predict than others. The discrepancy comes partly down to the scoring rates. Take hockey. Teams playing in the NHL score two or three goals per game on average. Compare that to basketball, where NBA teams will regularly score a hundred points in a game. If goals are rare—as they are in hockey—a successful shot will have more impact on the game. This means that a chance event, such as a deflection or lucky shot, is more likely to influence the final result. Low-scoring games also mean fewer data points to play with. When a brilliant team beats a lousy team 1–0, there is only one scoring event to analyze.

Fortunately, it's possible to squeeze extra information out of a game. One approach is to measure performance in other ways. In hockey, pundits often use stats such as the "Corsi rating"—the difference between the number of shots directed at an opponent's net and number targeted at a team's own goal—to make predictions about score lines. The reason they use such rating systems is that the number of goals scored in previous games does not say much about a team's future ability to score goals.

Scoring is far more common in games such as basketball, but the way in which the game is played can affect predictability, too. Haralabos Voulgaris has spent years betting almost exclusively on basketball and is now one of the world's top NBA bettors. At the MIT Sloan Sports Analytics Conference in 2013, he pointed out that the nature of scoring in basketball was changing, with players attempting more long-distance three-point shots. Because randomness plays a bigger role in these types of shots, it was becoming harder to predict which team would score more points. Traditional forecasting methods assume that team members work together to get the ball near the basket and score; these approaches are less accurate when individual players make speculative attempts from farther away.

Why does Voulgaris bet on basketball rather than another sport? It comes partly down to the simple fact that he likes the game. Sifting through reams of data doesn't make for a great lifestyle if it's not interesting. It also helps that Voulgaris has lots of data to sift through. Models need to process a certain amount of data before they can churn out reliable predictions. And in basketball plenty of information is available to analyze. The same cannot be said for other sports, however. In the early days of English soccer prediction, it was a struggle to dig up the necessary data. Whereas American pundits were dealing with a flood of information, in the United Kingdom there was barely a puddle. "We don't realise how easy we have it sometimes these days," Stuart Coles said.

With soccer data hard to come by in the late 1990s, gamblers had to get ahold of the information in any way they could. In some cases, people created automated programs that would scour the handful of websites that did publish results and copy the data tables straight off the webpages. Although this "screen scraping" provided a source of data, the websites being scraped did not like gamblers taking their content and clogging up their servers. Some installed countermeasures—such as blocking certain IP addresses—to stop people taking their data.

Even in the data-rich world of US sports, there is still plenty of variation in information between different leagues. One of the reasons Kent analyzed college sports was the amount of information available. "There are so many more games in college basketball, and a lot more teams," Kent said. "You get a huge database." Having access to these data helped Kent predict the outcomes of matches and place appropriate bets beforehand.

THROUGHOUT KENT'S CAREER, SPORTS betting in Las Vegas would come to a halt once a game started. By the time the referee blew the whistle to start the match, Kent's money was already down. The gambling and the action, two things that seemed to be so closely linked, were instead separated. It wasn't until 2009, when a new company arrived in the city, that casinos finally fixed this broken Venn diagram of gambling. That company was Cantor Gaming, part of the Wall Street firm Cantor Fitzgerald. In recent years, it has become the resident bookmaker at a number of major casinos. Walk into the sports section of the Venetian, the Cosmopolitan, or the Hard Rock, and you'll find dozens of big screens and betting machines, all operated by Cantor. Crammed between coverage of everything from baseball to football, there are rows of numbers and names, showing the odds for different matches. These "betting lines" rise and fall with the noise of the crowds. The room feels like a hybrid of a sports bar and a trading floor, a place where drinks and data blend together under the eternally bright casino lights.

The numbers on Cantor's screens might reflect the emotions of the spectators, but they are actually controlled by a computer program that adjusts the betting lines throughout the course of the game. Cantor calls it the "Midas algorithm." If something happens in the game, the program updates the odds on display automatically. Thanks to Midas, in-play betting has taken off in a big way in Las Vegas.

Much of the credit for the Midas software goes to an Englishman named Andrew Garrood, who joined Cantor in 2000. Before that, he'd been working as a trader for a Japanese investment bank. The leap to Las Vegas wasn't as big as it might seem: Garrood simply went from designing models that could price financial derivatives to ones that could put a value on sports results.

Cantor's biggest statement of intent came in 2008, when it bought a company called Las Vegas Sports Consultants. This company came up with odds for bookmakers across Nevada, including nearly half the casinos in Las Vegas. Cantor wasn't just interested in its predictions, however. In buying the company, Cantor had secured an extensive database of past results for a whole range of sports. The information would form a vital part of Cantor's analysis. From baseball to football, Cantor needed to know how certain events changed a game. If the San Francisco Giants get another home run, how would it affect their chances of winning? If the New England Patriots have one last attempt to get the ball down in the dying moments of a game, how likely are they to pull it off?

According to Garrood, straightforward "vanilla" events are relatively easy to predict. For instance, it's not too difficult to work out the chances a football team will score a touchdown if they start a drive from the 20-yard line. The problem is that there could be many successes and failures during a game, some of which are subtler than others. Which events are worth knowing about? Garrood has found that most plays don't affect the outcome much. It's therefore important to pin down the crucial events, the ones that do make a big difference. This is where the enormous database comes in handy. While many gamblers are relying on gut instinct, Midas can assess just how much effect that touchdown will really have.

How does Cantor make sure it gets all its predictions right? The answer is that it doesn't try to. There is a commonly held view that firms like Cantor use models to try to nail the correct result for every game. Matthew Holt, director of sports data at Cantor, has rebuffed this myth. "We don't make lines to predict the outcome of a game," he said in 2013. "We make lines in anticipation of where the action will come."

When it comes to betting, bookmakers' aims are fundamentally different from those of gamblers. Suppose two tennis players in the US Open are perfectly matched. The game is 50/50, which means that for a $1.00 bet, a fair return would be $1.00: if a gambler bet on both players, the bettor would come out even. But a bookmaker won't offer odds that return $1.00. Instead, it might offer a payoff of $0.95. Anyone who bets on both players will therefore end up $0.05 poorer.

If the same total amount is wagered on each player, the bookmaker will lock in a profit. But what if most bets go on one of the players? The bookmaker will need to adjust the odds to make sure it stands to gain the same amount regardless of who wins. The new odds might suggest one player is less likely to come out on top. Smart gamblers, who know that both players are equally good, will therefore bet on the one with longer odds. For bookmakers that have done their job properly, this isn't a concern. They don't move their betting lines to match the true chance of a result happening. They do it to balance the books.

Every day, the Midas algorithm combines computer predictions with real betting activity, tweaking the odds as wagers come in. It performs this juggling act for dozens of different sports, simultaneously updating betting lines as games progress. To make a profit, bookmakers like Cantor have to understand where gamblers' money is going. What are they betting on? How might they react to a particular event?

Just as information flows between bettors and bookmakers, in many instances gamblers will also try to work out what their rivals are doing. When word gets out that a betting syndicate has come up with a successful strategy, others inevitably want a piece of the action. Because many betting strategies have their origins in academia, it's often possible to piece together the basic models by sifting through research articles. But sports betting is a competitive industry, which means some of the most effective techniques remain shrouded in secrecy. According to sports statistician Ian McHale, "The proprietary nature of prediction models means that the published ones rarely (if ever) represent the very best models."

If gamblers don't know who has the best strategy, it can create a tense environment. In the huge Asian markets, where many of the largest soccer bets are placed, wagers are often made via instant messaging software. At the same time, information ricochets between bookmakers and gamblers, as each tries to work out what the other is thinking and how the other will bet. "The betting grapevine is huge," as one industry insider put it. "There is a lot of paranoia."

WHEN ASIAN BOOKMAKERS GET coverage in the Western media, it's not usually for a good reason. After some suspicious bowling in cricket matches between Pakistan and England in 2010, three of Pakistan's players were handed bans for agreeing to deliver bad balls. Reporters noted that bookmakers—many of whom were based in Asia—would often target such games. The scandals have since continued. During the summer of 2013, three cricketers playing in the Indian Premier League were charged with match fixing. Police claimed bookmakers had promised upward of $40,000 to the players if they let the opposition score runs at specific times. Then, in December 2013, UK police arrested six soccer players for allegedly offering to collect yellow or red cards to order.

There is certainly a huge appetite for gambling in Asia, and not all of it aboveboard. It's been estimated that the illegal betting market in China is ten times larger than the legitimate sums handled by the Hong Kong Jockey Club. Illegal betting is also common in India. When the national cricket team plays arch rival Pakistan, the total amount wagered can approach $3 billion. Yet the Asian betting market is changing. Gamblers no longer need to track down black market bookmakers in rooms behind backstreet bars. There was a time when they would need to bring cash and a code word; now they can bet by phone or online. Glossy call centers have replaced grimy betting rooms. The new industry is a step away from the illegal black market, but it remains little regulated. This is the "gray market": modern, corporate, and opaque.

When it comes to high-stakes betting on sports such as soccer, Asia is the location of choice for many Western gamblers. The reason is simple. In Europe and the United States, bookmakers rarely take large bets. As a result, gamblers based in these regions are finding it harder to stake the sort of money that will make their strategies profitable. Despite being a prolific bettor—or rather *because* he is a prolific bettor—Haralabos Voulgaris has complained that US bookmakers are reluctant to take his bets. Even when they do, the betting limits are placed at unhelpfully low levels; he might be allowed to stake only a few thousand dollars. Not all Western bookmakers shun successful bettors, however. In the past decade, one firm has gained a reputation for accepting—and even encouraging—bets from smart gamblers.

When Pinnacle Sports started in 1998, it was clear that it had some bold ambitions. The betting limits were high, with maximum stakes larger than those offered by many existing bookmakers. Pinnacle claimed it was happy to let players bet the maximum as often as they liked. Even if a player consistently made money, Pinnacle wouldn't shut the player down. Back in 2003, such ideas went completely against established bookmaking wisdom. If you want to make money, went the dogma, don't let smart bettors place huge bets. And certainly don't let them do it again and again. So, how did Pinnacle pull it off?

Whereas all bookmakers look at overall betting activity, Pinnacle also puts a lot of effort into understanding who is placing those bets. By accepting wagers from sharp bettors, Pinnacle can get an idea of what these gamblers think might happen. It's not very different from Bill Benter combining his predictions with the public odds displayed at Happy Valley. Sometimes the public knows things that a betting syndicate—or a bookmaker—might not.

Pinnacle generally posts an initial set of odds on Sunday night. It knows these numbers might not be perfect, so only take a small amount of bets at first. It has found that the first bets almost always come from talented small-stakes bettors: because the early odds are often incorrect, sharp gamblers pile in and exploit them. But Pinnacle is happy to hand an advantage to these so-called hundred-dollar geniuses if it means ending up with much better predictions about the games. In essence, Pinnacle pays smart gamblers for information.

The strategy of purchasing information has been attempted in other walks of life, too, sometimes with controversial results. In the summer of 2003, US Senators stumbled across a Department of Defense proposal for a "policy analysis market" that would allow traders to speculate on events in the Middle East. It would be possible to bet on events such as a biochemical attack, for example, or a coup d'état, or the assassination of an Arab leader. The idea was that if anyone had inside information and tried to exploit it, the Pentagon would be able to spot the change in market activity. Investors might make a profit, but they would also reveal their hand in the process. Robin Hanson, the economist behind the proposal, pointed out that intelligence agencies by definition pay people to report unsavory details. In moral terms, he didn't see a market as any better or worse than other types of transactions.

The Senators disagreed. One called the idea "grotesque"; another said it was "unbelievably stupid." According to Hillary Clinton, the policy would create "a market in death and destruction." The proposal did not survive long in the face of such fierce opposition. By the end of July, the Pentagon had scrapped the idea. The decision was arguably an ethical rather than economic one. Although critics attacked the morality of the proposal, few disputed that betting markets can reveal valuable insights about an event. Unlike participants in an opinion poll, gamblers have a financial incentive to be right.

When they make predictions about the future, they are putting their money where their mouth (or model) is.

Today, Pinnacle canvasses gamblers' opinions on a wide range of subjects. People can bet on the identity of the next president or who will take home an Academy Award. Pinnacle has so much faith in the approach that it regularly takes large bets on popular events: in the past, it has been possible to wager half a million dollars on the soccer Champions League final. Because Pinnacle's business model relies on having accurate predictions, there are some things it doesn't take bets on. In 2008, for instance, Pinnacle dropped horse racing as a betting option because it doesn't specialize in the sport.

Companies like Pinnacle, which have found a way to combine in-house statistical predictions with the opinions of smart gamblers, have challenged traditional bookmaking. By harnessing the knowledge of smart gamblers, they have more confidence in their odds, and hence are happy to take larger bets. Yet bookmakers are not the only ones changing. In some cases, gamblers are skipping the bookmaker altogether.

DURING THE PAST DECADE or so, approaches to betting have changed dramatically. As well as wagers moving online, bookmakers have faced competition from a new type of gambling market, in the form of the betting exchange. This is much like a stock exchange, except instead of buying and selling shares, gamblers can offer and accept wagers. Perhaps the best-known betting exchange is the London-based Betfair, which handles over seven million bets a day.

Betfair's creator, Andrew Black, came up with the idea for the website during the late 1990s, when he was a programmer at the British Government Communication Headquarters in Gloucestershire. Security wouldn't let him stay on site past five o'clock, so he found himself spending each evening alone in his rural farmhouse. Having so much free time was a burden, but it was also rather fruitful. "The boredom was horrendous," he later told the *Guardian*, "but mentally I became really quite productive."

While attending college, Black had developed an interest in betting. But there were some drawbacks with the traditional way of gambling, and during those evenings in Gloucestershire, Black thought about how things could be improved. Rather than going through a bookmaker, as he'd always had to do, why not let gamblers bet directly against each other? The project meant combining ideas from financial markets, gambling, and online retail. Black, who had previously spent time as a professional gambler, stock trader, and website builder, had experience in all three of these areas.

The Betfair website launched in 2000. That summer, the company arranged for a mock funeral procession to pass through the city of London, with a coffin announcing the "death of the bookmaker." Although the stunt brought plenty of media coverage, competitors were already lurking. One rival website mimicked eBay: if someone wanted to place a bet of £1,000 at certain odds, the site would try to pair the bettor up with someone happy to take such a bet. Trying to pair people up was a bit like playing a giant online game of snap. And that sometimes meant waiting a long time for a match.

Fortunately, Betfair had a way to speed things up. If there were no takers for a wager, the website would divide up the bet between several different people. Rather than trying to find someone willing to take the full £1,000, for example, the website might slice up the total and match it with, say, five people wanting to accept a £200 wager. Whereas bookmakers had traditionally made their money by tweaking the odds on display, Betfair left the odds untouched and instead took a cut from the profits of whoever won a particular bet.

BETTING EXCHANGES LIKE BETFAIR have opened up a new approach to gambling. Unlike traditional bookmakers, you aren't limited to betting for a particular result. You can also "lay" the result by taking the other side of the bet; if the result doesn't happen, you win whatever was staked.

Because you can bet both ways on a betting exchange, it's possible to make money before a match ends. Suppose a betting exchange currently displays odds of 5 for a particular team. You decide to bet £10 on them, which means you'll get back £50 if they win. Then something changes. Perhaps the opposition's star player picks up an injury. Your team is more likely to win now, so the odds drop to 2. Rather than waiting until the match ends—and risking the result going against you—you can hedge your original bet by accepting someone else's £10 bet at the lower odds. If your team wins, you'll get £50 from the first bet but have to cough up £20 for the second; if your team loses, the two bets will cancel each other out, as shown in Table 4.1. The match hasn't even started and you're guaranteed £30 if your team wins, and you won't lose anything if they don't. (Many bookmakers have since introduced a "cashout" feature, which in essence reproduces these trades.)

Because you can back and lay each result, the Betfair website displays two columns for every match, showing the best available odds on each side of the bet. Such technology has made it easier for gamblers to see what others are thinking and to take advantage of odds they believe are incorrect. Yet it's not just bookmaking that is becoming more accessible.

TABLE 4.1. Bets Can Be Hedged by Backing and Then Laying the Same Team

		1st bet	2nd bet	Total
Result	Your team wins	£50	–£20	£30
	Your team loses	–£10	£10	£0

SCIENTIFIC BETTING STRATEGIES HAVE traditionally been the preserve of private betting syndicates like the Computer Group or, more recently, consultancy firms like Atass. This may not be the case for much longer. Just as banks offer clients access to investment funds, some companies are letting people invest in scientific gambling methods. As Bloomberg columnist Matthew Klein put it, "If I find a guy who is good at sports betting and is willing to bet with my money in exchange for a fee, he is, for all intents and purposes, a hedge fund manager." Rather than putting money into established asset classes such as shares or commodities, investors now have the option of sports betting as an alternative asset class.

Betting might seem somewhat distant from other types of investment, but that is one of its selling points. During the 2008 financial crisis, many asset prices fell sharply. Investors often try to build a diverse portfolio to protect against such shocks; for example, they might hold stocks in several different companies in a range of industries. But when markets run into trouble, this diversity is not always enough. According to Tobias Preis, a researcher in complex systems at the University of Warwick, stocks can behave in a similar way when a financial market hits a rough period. Preis and colleagues analyzed share prices in the Dow Jones Industrial Average between 1939 and 2010 and found that stocks would go down together as the market came under more stress. "The diversification effect which should protect a portfolio melts away in times of market losses," they noted, "just when it would most urgently be needed."

The problem isn't limited to stocks. In the run-up to the 2008 crisis, more and more investors began to trade "collateralized debt obligations." These financial products gathered together outstanding loans such as home mortgages, making it possible for investors to earn money by taking on some of the lenders' risk. Although there might have been a high probability that a single person would default on a loan, investors assumed it was extremely unlikely everyone would default at the same time. Unfortunately, this assumption turned out to be incorrect. When one home lost its value during the crisis, others followed.

Advocates of sports betting point out that wagers are generally unaffected by the financial world. Games will still go ahead if the stock market takes a dive; betting exchanges will still accept wagers. A hedge fund that concentrates on sports betting should therefore be an attractive investment, because it provides diversification. It was this idea that persuaded Brendan Poots to set up a sports-focused hedge fund in 2010. Based in Melbourne, Australia, Priomha Capital aimed to give the public investors access to the traditionally private world of sports prediction.

Creating good forecasts can require additional expertise, so Priomha linked up with researchers at the Royal Melbourne Institute of Technology. To some extent, the approach is a twenty-first-century version of the Computer Group's strategy. Priomha creates a model for a particular sport, runs simulations to predict the likelihood of each result, and then compares the predictions with the current odds on betting exchanges such as Betfair.

The big difference is that investors are not restricted to betting before a game starts. Which is good news, because Poots has found that odds generally settle down to a fair value in the run-up to a fixture. "Come kick off, the market's pretty efficient," he said. "But once play starts, that's where we have a huge opportunity."

When it comes to soccer prediction, "in-play" analysis was always the natural next step. After working on final score predictions in 1997, Mark Dixon turned his attention to what happens during a soccer match. Along with fellow statistician Michael Robinson, he simulated matches using a similar model to the one he'd published with Stuart Coles, but with some important new modifications. As well as accounting for each teams' attacking strength and defensive weakness, the model included factors based on the current score and time left to play. It turned out that including in-play information led to more accurate predictions than the original Dixon-Coles model.

The model also made it possible to test popular soccer "wisdom." Dixon and Robinson noted that commentators would often tell viewers that teams were more vulnerable after they scored a goal. The researchers referred to this cliché as "immediate strike back." The idea is that after a goal goes in, attackers' concentration wobbles, which can allow the opposition back into the game. But the cliché turned out to be misguided. Dixon and Robinson found that teams weren't especially vulnerable after they have scored a goal. So, why did commentators often claim that they were?

If we come across something unusual or shocking, it stands out in our mind. According to Dixon and Robinson, "People have a tendency to overestimate the frequency of surprising events." This doesn't just happen in sports. Many worry more about terrorist attacks than bathtub accidents, despite the fact that—in the United States at least—you're far more likely to die in a bathtub than at the hands of a terrorist. Unusual events are more memorable, which also explains why people think it's easier to become a millionaire with a one-dollar lottery ticket than by playing roulette repeatedly. Although both are terrible ideas, in terms of raw probability, playing roulette again and again is more likely to generate a lucky million dollars in profit.

Betting successfully during a soccer match means identifying human biases like these. Are there certain aspects of the game that gamblers consistently misjudge? Poots has found that a few things stand out. One is the effect of goals. Just as Dixon and Robinson noted, the popular view is not always the correct one: a goal doesn't always create the shock people think that it does. Gamblers also tend to overestimate the impact of red cards. That isn't to say they don't have any effect. A team playing against an opponent with ten men will probably score at a higher rate (one 2014 study reckoned the rate would be 60 percent higher on average). But the odds often move too far, suggesting gamblers mistake a difficult situation for a hopeless one.

Following a dramatic event, the odds available on a betting exchange gradually adjust to the new situation. When things have settled down, Priomha can hedge its bets by taking the opposite position. If it backed the home team to win at long odds, perhaps after a red card, it will bet against them when the odds decrease. This way, it doesn't matter how the game ends. Like a trader who buys an item from a panicked seller and later sells it back at a higher price, the team closed their position and offloaded any remaining risk.

There are plenty of opportunities to capitalize on inaccurate odds during a game. Unfortunately, there are also fewer bets on offer, which means Priomha has to be careful not to disrupt the market with large wagers. "During play, you have to drip feed your money in," Poots said. In fact, the size of the market is one of the biggest obstacles facing funds like Priomha. Because it makes money by identifying incorrect sports odds, the more money it has to invest, the more erroneous odds it has to find.

The current plan is to manage up to $20 million of investors' money. Poots pointed out that if they tried to handle a much larger amount—such as $100 million—it would be a struggle to make reasonable returns. They might be able to find enough opportunities to bring in a 5 percent annual return, but as a hedge fund, they really want to be making double figures for their investors, and they are more likely to achieve this if they constrain the size of the fund.

Although Priomha has not reached its limit yet, as the fund grows Poots is noticing a change in who is buying into the strategy. "Our investor profile used to be someone who likes sport, and liked to have a bet," he said. "It's now becoming people who have got their pension or other funds to invest."

Priomha is not the only sports betting fund to have appeared in recent years. The London-based Fidens syndicate opened its fund to investors in 2013; two years later it was managing more than £5 million. Mathematics graduate Will Wilde heads up Fidens's trading strategy. This involves betting on ten soccer leagues around the globe, placing around three thousand wagers per year.

Stock market investing has often been compared to gambling, especially when shares are held for only a short period of time. There is a certain irony then, that gambling is increasingly seen as a viable option for investors. Not all sports betting funds have been successful, however. In 2010, investment firm Centaur launched the Galileo fund, which was designed to allow investors to profit from sports betting. The plan was to attract $100 million of investment and generate an annual return of 15 to 25 percent. The finance community watched with interest, but two years later the fund folded.

Although the ambitions of funds like Priomha are currently constrained by the size of the betting market, things could be very different if sports betting were to expand in the United States. "If America was to open up," Poots said, "the whole game changes completely." The first major hints of change came soon after Priomha was founded. Following a referendum in 2011, New Jersey governor Chris Christie signed a bill legalizing sports betting in the state. For the first time, gamblers in Atlantic City would be able to bet on games like the Super Bowl. That was the theory at least. It did not take long for professional sports leagues to bring in lawyers to halt the expansion. The case has been bouncing through the court system ever since, the main obstacle being a federal law from 1992 that prohibits sports betting in all but four states. Opponents say gambling should be limited to places like Las Vegas; New Jersey claims the law is unconstitutional and that the public supports legalized betting. Indeed, many sports leagues already allow people to put money on their predictions coming true. Every year people pay to take part in fantasy sports leagues, even though betting on a specific match outcome is still illegal.

Advocates for law changes say there are two main advantages to legalized gambling. First, it would generate more tax. It's been estimated that less than 1 percent of sports bets in the United States are placed legally. The remaining 99 percent of wagers, made through unlicensed bookmakers or offshore websites, probably run into hundreds of billions of dollars. If these bets were legal, the tax revenue would be enormous. Second, legalization means regulation, and regulation means transparency. Bookmakers and betting exchanges keep records of customers, and online firms also have bank details. According to NBA commissioner Adam Silver, legalizing gambling would bring the activity into view of government scrutiny. "I believe that sports betting should be brought out of the underground and into the sunlight where it can be appropriately monitored and regulated," he wrote in the *New York Times* in 2014.

Betting syndicates would also stand to benefit from legalization. With more bookmakers taking wagers, syndicates could place bets on a much grander scale. There is also a chance that new laws will allow syndicates to bet in Las Vegas. Currently, if gamblers want to bet on sports in the city, they still need to turn up at a casino with a handful of cash, which makes it difficult to systematically place large bets. In 2015, the Nevada senate passed a bill that would allow a group of investors to back a bettor, which is essentially what Priomha already does outside of the United States. If the bill gets through the state assembly and becomes law, many more sports hedge funds could appear. Other countries are also debating new gambling laws. In Japan, sports bettors can currently put money only on horse, boat, or cycle races. A new bill, submitted in April 2015, and supported by the prime minister, proposes to change that. New opportunities will also arise in India and China, as informal betting markets become more regulated.

According to sports journalist Chad Millman, it is not just established gamblers who would be well positioned to profit from law changes. During a visit to MIT in March 2013, Millman got talking to Mike Wohl, an MBA student at the university's business school. For his study project, Wohl had considered gambling as "the missing asset class." Wohl had a background in finance, and his analysis—along with his personal experience of betting—suggested that sports wagers could produce as good a trade-off between risk and return as investing in stocks could.

Millman pointed out that there are two extremes to the gambling spectrum. At one end are professional sports bettors, the so-called sharps who regularly place successful bets. At the other are the everyday gamblers, who don't have predictive tools or reliable strategies. In between, Millman says, are a number of people like Wohl who have the necessary skills to bet successfully but haven't yet chosen to use them. They might work in finance or research; perhaps they have MBAs or PhDs. If sports betting was to expand in the United States, these small-scale bettors would be in a good position to profit. With their quantitative backgrounds, they are already familiar with the crucial methods. They also have the necessary tools, thanks to increases in computing power and data availability. All they need now is the access.

THERE ARE CERTAIN ADVANTAGES to being a betting start-up. For one, it means more flexibility. But should new syndicates follow sports betting strategies that have already been successful? Or should they exploit their flexibility and try something else?

In retrospect, Michael Kent would look at matches in far more detail. "If I was starting over right now," he said, "I would want to have play-by-play data." The additional information would make it possible to measure individual contributions. This would be a stark contrast with his previous analysis: in his models, Kent has always treated teams as a single entity. "I have no knowledge of players," he said. "I know what the team did, but I don't know the name of the quarterback."

Some modern betting syndicates go to great lengths to measure individual performances. "We do analysis on the effect of every player in every team," Will Wilde said. "Every player has a rating that goes up or down, regardless of whether they play or not." In Hong Kong, Bill Benter's syndicate even employs people to sift through videos of races. They might look at how a horse's speed changes during the race or how well it recovers after a bump. These "video variables" make up a relatively small part of the model—about 3 percent—but they all help nudge the predictions a little closer to reality.

It is not always just a matter of collecting more data. In soccer, successful defenders can be a nightmare for statisticians. During his years playing for Milan and Italy, Paolo Maldini averaged one tackle every other game. It wasn't because he was a lazy player; it was because he didn't need to make many tackles. He held back the opposition by getting into the right positions. Raw statistics such as number of tackles can therefore be misleading. If a defender makes fewer tackles, it doesn't always mean he's getting worse. It could mean he's improving.

A similar problem crops up with cornerbacks in US football. Their job is to patrol the edges of the field, defending against attacking passes by the opposition. Good cornerbacks intercept lots of passes, but great ones won't need to: the other team will be trying to avoid them. As a result, the best cornerbacks in the NFL might touch the ball only a handful of times per season.

How can we measure a player's ability if they rarely do anything that can be measured? One option is to compare the overall team performance when a player is and isn't on the field. At the simplest level, we could look at how often a team wins when a certain individual is playing. Sometimes it is clear that a player is valuable to a team. For example, when striker Thierry Henry played for Arsenal soccer club between 1999 and 2007, the team won 61 percent of matches he appeared in. On the other hand, they won only 52 percent of the games he missed.

Counting wins is simple enough, but measuring players in this way can raise some unexpected results. In some cases, it might even appear that fan favorites are not actually that important to the team. Since Steven Gerrard made his first appearance for Liverpool in 1998, they have won half the games he's played in. Yet they have also won half the games he hasn't had a role in. Brendan Poots points out that the best clubs have strong squads, so can often cope with losing a star individual. When top players go off injured, teams adjust. "In the sum of the parts," Poots said, "the effect that they have—or their absence has—is not as great as people think."

The problem with simply tallying up wins with and without a certain player, however, is that the calculation doesn't account for the importance of those games or the strength of the opposition. Teams often field more big-name players in crucial matches, for example. One way to get around these issues is to use a predictive model. Sports statisticians often assess the importance of a particular player by comparing the predicted scores for the games the player played in with the actual results of those games. If the team performs better than expected when that player is on the field, it suggests the player is especially important to the team.

Again, it's not always the best-known players who come out on top. This is because identifying the most important player is not the same as finding the best player. The most important player—as judged by the model—might be someone without an obvious replacement or a player whose style suits the team particularly well.

To interpret the results of their predictive models, firms working on sports forecasts employ analysts with a detailed knowledge of each team. These experts can suggest why a certain player appears to be so important and what that might mean for upcoming matches. Such information is not always easy to quantify, but it might have a big effect on results. The trick is to know what the model doesn't capture and to account for such features when making predictions. Sports statistician David Hastie points out that this goes against many people's idea of a scientific betting strategy. "There is a common perception that betting is all about models," he said. "People expect a magic formula."

GAMBLERS NEED TO KNOW how to get at crucial information, whether it is quantitative, as is the case with model predictions, or of a more qualitative nature, as with human insights. Although well known for his computer models, Kent knew the importance of human experts when making predictions. He received regular updates from people with in-depth knowledge of certain sports, people whose job it was to know things that the model might not capture. "We had a guy in New York City who could tell you the starting lineup for 200 college basketball teams," he said.

Making better predictions about individual players doesn't just benefit gamblers. As techniques improve, bettors and sports teams are finding more common ground, drawn together by a common desire to anticipate what will happen in the next season, or the next game, or even the next quarter. Every spring, team managers chat with statisticians and modelers at the MIT Sloan Sports Analytics Conference. Prediction methods can be particularly useful when teams go scouting for new signings. Historically, assessing a player's value has been difficult because performances are subject to chance. A player might have an impressive—and lucky—season one year, and then have a less successful time the next.

The "*Sports Illustrated* jinx" is a well-known example of this problem: often a player who appears on the cover of *Sports Illustrated* subsequently suffers a dip in form. Statisticians have pointed out that the *Sports Illustrated* jinx is not really a jinx. Players who end up on the cover often do so because they've had an unusually good season, which was down to random variation rather than a reflection of their true ability. The drop in performance that came the following year was simply a case of regression to the mean, just as Francis Galton found while studying inheritance.

When a club signs a new player, it has to make decisions based on past accomplishments. Yet what it is really paying for is future performances. How can a sports club predict a player's true ability? Ideally, it would be possible to pull past performances apart and work out how much they were influenced by ability and by chance. Statistician James Albert has attempted to do this for baseball. By trawling through lots of different statistics for pitchers, including wins and losses, strikeouts—where the batter misses the ball three times—and runs scored against them. He found that the number of strikeouts was the most accurate representation of a pitcher's true skill, whereas statistics such as home runs conceded were more subject to chance, and hence are a poor reflection of pitching ability.

Other sports are trickier to analyze. Soccer pundits generally use simple measurements, such as goals per game, to quantify how good strikers are. But what if strikers play for a good team and benefit from having other players setting them up with scoring chances? In 2014, researchers at Smartodds and the University of Salford assessed the goal-scoring ability of different soccer players. Rather than just asking how likely a striker was to score, they split goal scoring into two components: the process of generating a shot—which could be influenced by the team performance—and the process of converting that shot into a goal. Splitting up scoring in this way led to far better predictions about future goal tallies than simple goals-per-game statistics provided. The study also produced some unexpected conclusions. For instance, it appeared that the number of shots a player had bore little relation to the team's attacking ability. In other words, good players generally end up with a similar shot count regardless of whether they are playing for a great team or a weak one. Although better teams have more shots overall, a decent player ends up being a little fish in a large scoring pool; at a struggling club, that same player can make a bigger contribution to the total. The researchers also found that it was difficult to predict how often a player would convert shots into goals. Hence, they suggest that team managers looking at a potential signing should estimate how many shots that player generates rather than how many goals the player scores.

WHEN IT COMES TO scientific sports betting, the most successful gamblers are often the ones who study games others have neglected. From Michael Kent's work on college football to Mark Dixon and Stuart Coles's research in soccer, the big money generally comes from moving away from what everyone is doing.

Over time, bookmakers and gamblers have gradually latched on to the best-known strategies. As a result, it is becoming harder to profit from major sports leagues. Erroneous odds are less common, and competitors are quick to jump on any advantage. New syndicates are therefore better off focusing their attention on lesser-known sports, where scientific ideas have often been ignored. According to Haralabos Voulgaris, this is where the biggest opportunities lie. "I would start with the minor sports," he said at the MIT Sloan Sports Analytics Conference in 2013. "College basketball, golf, NASCAR, tennis."

In minority sports, additional knowledge—whether from models or experts—can prove extremely valuable. Because crucial variables are not so well known, the difference in skill between a sharp bettor and a casual gambler can be huge. As well as helping gamblers build better predictive models, improvements in technology are also changing how bets are made. The days of suitcases full of banknotes are coming to an end. Bets can be placed online, and gamblers can control hundreds of wagers at the same time. This technology has also paved the way for new types of strategy. A large part of sports betting throughout its history has been about forecasting the correct result. But scientific betting is no longer just a matter of predicting score lines. In some cases, it is becoming possible to know nothing about the result and still make money.

5

RISE OF THE ROBOTS

"WHAT HATH GOD WROUGHT!" THE MESSAGE READ. IT WAS May 24, 1844, and the world's first long-distance telegram had just arrived in Baltimore. Thanks to Samuel Morse's new telegraph machine, the biblical quote had traveled along a wire all the way from Washington, DC. Over the next few years, single-wire telegraph systems spread around the globe, creeping into the heart of all sorts of industries. Railway companies used them to send signals between stations, while police fired off telegrams to get ahead of fleeing criminals. It wasn't long before British financiers got ahold of the telegraph, too, and realized that it could be a new way to make money.

At the time, stock exchanges in the United Kingdom operated independently in each region. This meant there were occasional differences in prices. For example, it was sometimes possible to buy a stock for one price in London and sell it for a higher price in one of the provinces. Obtain such information quickly enough, and there was a profit to be made. During the 1850s, traders used telegrams to tell each other about discrepancies, cashing in on the difference before the price changed. From 1866 onward, America and Europe were linked by a transatlantic cable, which meant traders were able to spot incorrect prices even faster. The messages that traveled down the wire were to become an important part of finance (even today, traders refer to the GBP/USD exchange rate as "cable").

The invention of the telegraph meant that if prices were out of line in two locations, traders had the means to take advantage of the situation by buying at the cheaper price and selling at the higher one. In economics, the technique is known as "arbitrage." Even before the invention of the telegraph, so-called arbitrageurs had been on the hunt for mismatched prices. In the seventeenth century, English goldsmiths would melt down silver coins if the price of silver climbed past the value of the coin. Some would even trek further afield, hauling gold from London to Amsterdam to capitalize on differences in the rate of exchange.

Arbitrage can also work in gambling. Bookmakers and betting exchanges are merely different markets trading the same thing. They all have varying levels of betting activity and contrasting opinions about what might happen, which means their odds won't necessarily line up. The trick is to find a combination of bets so that whatever happens, the payoff will be positive. Suppose you're watching a tennis match between Rafael Nadal and Novak Djokovic. If one bookmaker is offering odds of 2.1 on Nadal, and another is offering 2.1 on Djokovic, betting $100 on each player will net you $210—and cost you $100—whatever the result. Whoever wins, you walk away with a profit of $10.

Unlike syndicates working on sports prediction, which are in essence betting that their forecast is closer to the truth than the odds suggest, arbitrageurs don't need to take a view on what will happen.

Whatever the result, the strategy should lead to a guaranteed profit, so long as a gambler can spot the opportunity in the first place. But how common are arbitrage situations?

In 2008, researchers at Athens University looked at bookmakers' odds on 12,420 different soccer matches in Europe and found 63 arbitrage opportunities. Most of the discrepancies occurred during competitions such as the European Championship. This was not particularly surprising, because tournament results are generally more variable than results in leagues where teams play each other often.

The following year, a group at the University of Zurich searched for potential arbitrage in odds given by betting exchanges like Betfair as well as traditional bookmakers. When they considered both types of market, there were far more stray odds. They found it would have been possible to make a guaranteed profit on almost a quarter of games. The average return wasn't huge—around 1 to 2 percent per game—but it was clear there were enough inconsistencies to make arbitrage a viable option.

Despite the allure of arbitrage betting, there are some potential pitfalls. To be successful, gamblers need to set up accounts with a large number of bookmakers. These companies usually make it easy to deposit money but hard to withdraw it. Bets also need to be placed simultaneously: if one wager lags behind another, the odds might change, thwarting any chance of a guaranteed profit. Even if gamblers can overcome these logistical issues, they have to avoid attracting the attention of the bookmakers themselves, who generally dislike having arbitrageurs cutting into their profits.

It is not just differences between bookmakers that can be exploited. Economist Milton Friedman pointed out that there is a paradox when it comes to trading. Markets need arbitrageurs to take advantage of incorrect prices and make them more efficient. Yet, by definition, an efficient market shouldn't be exploitable, and hence shouldn't attract arbitrageurs. How can we explain this contradictory situation? In reality, it turns out that markets often have short-term inefficiencies. There are periods of time when prices (or betting odds) do not reflect what is really going on. Although the information is out there, it hasn't been processed properly yet.

After a major event—such as a goal being scored—gamblers on betting exchanges need to update their opinions on what the odds should be. During this period of uncertainty, whoever reacts to the news first will be able to place bets against opponents who have not yet adjusted their odds. There is a limited window in which to do this. Over time the market will become more efficient, and the available odds will change to reflect the new information. In 2008, a group of researchers at the University of Lancaster reported that it takes less than sixty seconds for gamblers on betting exchanges to adjust to a dramatic event in a football match.

Not only is the betting window small, potential gains can be modest, too. To profit a gambler would need to place a large number of bets, and place them quickly. Unfortunately, this is not something that humans are particularly good at. We take time to process information. We hesitate. We struggle with multiple tasks. As a result, some gamblers are choosing to step back from the bustle of hectic betting markets. Where humans falter, the robots are rising.

THERE ARE TWO WAYS to access the Betfair betting exchange. Most people simply go to the website, which displays the latest odds as they become available. But there is another option. Gamblers can also bypass the website and link their computers directly to the exchange. This makes it possible to write computer programs that can place bets automatically. These robot gamblers have plenty of advantages over humans: they are faster, more focused, and they can bet on dozens of games at once. The speed of betting exchanges also works in their favor. Betfair is quick to pair up people who want to bet for a particular event with those intending to bet against it. Of the 4.4 million bets placed on the day of England's opening match in the 2006 soccer World Cup, all but twenty were handled in less than a second.

Automated gamblers are increasingly common in betting. According to sports analyst David Hastie, there are plenty of bots out there searching for stray odds and exploiting other gamblers' mistakes. "These algorithms mop up any mispricing," he said. The presence of artificial arbitrageurs makes it difficult for humans to cash in on such opportunities. Even if they spot an erroneous price, it's often too late to do anything about it. The bots will already be placing bets, removing these slices of profit from the market.

Arbitrage algorithms are also becoming popular in finance. As in betting, the faster the better. Companies are doing all they can to ensure they get to the action before their competitors do. It has led to many firms placing their computers directly next to stock exchange servers. When the market reacts quickly, even a slightly longer wire can lead to a critical delay in making a trade.

Some are going to even more extreme lengths. In 2011, US firm Hibernia Atlantic started work on a new $300 million transatlantic cable, which will allow data to cross the ocean faster than ever before. Unlike previous wires, it will be directly below the flight path from New York to London, the shortest possible route between the cities. It currently takes 65 milliseconds for messages to travel the Atlantic; the new cable aims to cut that down to 59. To give a sense of the scales involved, one blink of the human eye takes 300 milliseconds.

Fast trading algorithms are helping firms learn about new events first and act on them before others do. Yet not all bots are chasing arbitrage opportunities. In fact, some have the opposite aim. While arbitrage algorithms are searching for lucrative information, other bots are trying to conceal it.

WHEN SYNDICATES BET ON horse races in Hong Kong, they know that the odds will change after they've placed their bets. This is because in pari-mutuel betting the odds depend on the size of the betting pool. Teams therefore have to account for the shift when developing a betting strategy. If they put down too much, and shift the odds too far, they might end up worse off than if they'd bet less.

The problem also appears in sports betting. If you try to put down a large amount of money on a football match, it will be the bookmakers—or betting exchange users—who move the odds against you. Let's say you want to bet $500,000 on a certain outcome. One bookmaker might offer you odds that would return double the stake. But the bookmaker might only be willing to take a bet of $100,000 at those odds. After you place that initial bet, the bookmaker's odds will probably drop. Which means you've still got $400,000 you want to bet, and you've already disrupted the market. So, if you bet another $100,000 at the new odds, you won't quite double your money. You might get even lower odds for the next chunk of cash, and things will continue to get worse with each bet you make.

Traders call the problem "slippage." Although the price initially on offer might look good, it can slip to a less favorable price as the transaction is made. How can you get around the problem? Well, you could try to hunt down a bookmaker who'll take the bet in one go. At best, this could take a while; at worst, you'll never find one. Alternatively, you could place the first bet of $100,000 and then wait and hope the bookmaker's odds will rise again so you can bet the next chunk of money. Which is clearly not the most reliable strategy either.

A better approach would be to mimic the tactics employed by betting exchanges. Betfair's early success was in part the result of the way it handled each bet. Rather than attempting to find a gambler who wanted to accept a bet of the exact same size, Betfair sliced up the bet into smaller chunks. It was far easier—and quicker—to find several users happy to take on these little wagers than to hunt down a single gambler willing to take the whole bet.

The same idea makes it possible to sneak a trade into the market with limited slippage. Instead of trying to offload the whole trade at once, so-called order-routing algorithms can slice the main trade up into a series of smaller "child" orders, which can easily be completed. For the process to work effectively, algorithms need to have good knowledge of the market. As well as having information on who's happy to take the other side of each trade—and at what price—the program has to time the transactions carefully to reduce the chances of the market moving before the trade is complete. The resulting trade is known as an "iceberg order": although competitors see small amounts of trading activity, they never know what the full transaction looks like. After all, traders don't want rivals shifting prices because they know a big order is about to arrive. Nor do they want others to know what their trading strategy is.

Because such information is valuable, some competitors employ programs that can search for iceberg trades. One example is a "sniffing algorithm," which make lots of little trades to try to detect the presence of big orders. After the sniffing program submits each trade, it measures how quickly it takes to get snapped up in the market. If there's a big order lurking somewhere, the trades might go through faster. It's a bit like dropping coins into a well and listening for the splashes to work out how deep it is.

Although bots allow gamblers and banks to carry out multiple transactions quickly, they do not always act in the interests of their owners. Left unsupervised, bots can behave in unexpected ways. And sometimes they wander deep into trouble.

BY THE TIME THE 2011 Christmas Hurdle at Dublin's Leopardstown Racecourse reached the halfway mark, the race was as good as won.

It was just after two o'clock, and the horse named Voler La Vedette was already leading by a good distance. As the hooves pounded the ground on that cold December afternoon, nobody with any sense would have bet against that horse.

Yet somebody did. Even as Voler La Vedette approached the line, the Betfair online market was displaying extremely favorable odds for the horse that was almost certain to win. It appeared that someone was happy to accept bets at odds of 28: for every £1 bet, the bettor was offering to pay £28 if the horse won. Very happy, in fact. This remarkably pessimistic gambler was offering to accept £21 million worth of bets. If Voler La Vedette came first, the gambler would be on the hook for almost £600 million.

Soon after the race finished, one Betfair user posted a message on the website's forum. Having witnessed the whole bizarre situation, the user joked that someone must have been giving bettors a Christmas bonus. Others chipped in with potential explanations for the mishap. Maybe a gambler had suffered an attack of "fat fingers" and hit the wrong number on the keyboard?

It didn't take long for another user to suggest what might really have been going on. The person had noticed something odd about that offer to match £21 million of bets. To be precise, the number displayed on the exchange was just under £21.5 million. The user pointed out that computer programs often store binary data in units that contain thirty-two values, known as "bits." So, if the rogue gambler had designed a 32-bit program to bet automatically, the largest positive number the bot would be able to input on the exchange would be 2,147,483,648 pence. Which meant that if the bot had been doubling up its bets—just as misguided Parisian gamblers used to do while betting on roulette in the eighteenth century—£21.5 million is the highest it would have been able to go.

It turned out to be a superb piece of detective work. Two days later Betfair admitted that the error had indeed been caused by a faulty bot. "Due to a technical glitch within the core exchange database," they said, "one of the bets evaded the prevention system and was shown on the site." Apparently, the bot's owner had less than £1,000 in an account at the time, so as well as fixing the glitch, Betfair voided the bets that had been made.

As several Betfair users had already pointed out, such ridiculous odds should never have been available. The two hundred or so gamblers who had bet on the race would therefore have struggled to persuade a lawyer to take their case. "You cannot win—or lose—what is not there in the first place," Greg Wood, the *Guardian*'s racing correspondent, wrote at the time, "and even the most opportunistic ambulance-chaser is likely to take one look at this fact and point to the door."

Unfortunately, the damage created by bots isn't always so limited. Computer trading software is also becoming popular in finance, where the stakes can be much higher. Six months after the Voler La Vedette bot got its odds wrong, one financial company was to discover just how expensive a troublesome program could be.

THE SUMMER OF 2012 was a busy time for Knight Capital. The New Jersey–based stockbroker was getting its computer systems ready for the launch of the New York Stock Exchange's Retail Liquidity Program on August 1. The idea of the liquidity program was to make it cheaper for customers to carry out large stock trades. The trades themselves would be executed by brokers like Knight, which would provide the bridge between the customer and the market.

Knight used a piece of software called SMARS to handle customers' trades. The software was a high-speed order router: when a trade request came in from a client, SMARS would execute a series of smaller child orders until the original request had been filled. To avoid overshooting the required value, the program kept a tally of how many child orders had been completed and how much of the original request still needed to be executed.

Until 2003, a program named Power Peg had been responsible for halting trading once the order had been met. In 2005, this program was phased out. Knight disabled the Power Peg code and installed the tally counter into a different part of the SMARS software. But, according to a subsequent US government report, Knight did not check what would happen if the Power Peg program was accidently triggered again.

Toward the end of July 2012, technicians at Knight Capital started to update the software on each of the company's servers. Over a series of days, they installed the new computer code on seven of the eight servers. However, they reportedly failed to add it to the eighth server, which still contained the old Power Peg program.

Launch day arrived, and trade orders started coming in from customers and other brokers. Although Knight's seven updated servers did their job properly, the eighth was unaware of how many requests had already been completed. It therefore did its own thing, peppering the market with millions of orders and buying and selling stocks in a rapid-fire trading spree. As the erroneous orders piled up, the tangle of trades that would later have to be unraveled grew larger and larger. While technology staff worked to identify the problem, the company's portfolio grew. Over the course of forty-five minutes, Knight bought around $3.5 billion worth of stocks and sold over $3 billion. When it eventually stopped the algorithm and unwound the trades, the error would cost it over $460 million, equivalent to a loss of $170,000 per second. The incident left a massive dent in Knight's finances, and in December of that year the company was acquired by a rival trading firm.

Although Knight's losses came from the unanticipated behavior of a computer program, technical problems are not the only enemy of algorithmic strategies. Even when automated software is working as planned, companies can still be vulnerable. If their program is too well behaved—and hence too predictable—a competitor might find a way to take advantage of it.

In 2007, a trader named Svend Egil Larsen noticed that the algorithms of one US-based broker would always respond to certain trades in the same way. No matter how many stocks were bought, the broker's software would raise the price in a similar manner. Larsen, who was based in Norway, realized that he could nudge up the price by making lots of little purchases, and then sell a large amount of stock back at the higher price. He'd become the financial equivalent of Professor Pavlov, ringing his bell and watching the algorithm respond obediently. Over the course of a few months, the tactic earned Larsen over $50,000.

Not everybody appreciated the ingenuity of his strategy. In 2010, Larsen and fellow trader Peder Veiby—who'd been doing the same thing—were charged with manipulating the market. The courts seized their profits and handed the pair suspended sentences. When the verdict was announced, Veiby's lawyer argued that the nature of the opponent had biased the ruling. Had the pair profited from a stupid human trader rather than a stupid algorithm, the court would not have reached the same conclusion. Public opinion sided with Larsen and Veiby, with the press comparing their exploits to those of Robin Hood. Their support was vindicated two years later, when the Supreme Court overturned the verdict, clearing the two men of all charges.

There are several ways algorithms can wander into dangerous territory. They might be influenced by an error in the code, or they might be running on an out-of-date system. Sometimes they take a wrong turn; sometimes a competitor leads them astray. But so far we have only looked at single events. Larsen targeted a specific broker. Knight was a lone company. Just one gambler offered ridiculous odds on Voler La Vedette. Yet there are an increasing number of algorithms in betting and finance. If a single bot can take the wrong path, what happens when lots of firms use these programs?

DOYNE FARMER'S WORK ON prediction did not end with the path of a casino roulette ball. After obtaining his PhD from UCLA in 1981, Farmer moved to the Santa Fe Institute in New Mexico. While there, he developed an interest in finance. Over a few short years, he went from forecasting roulette spins to anticipating the behavior of stock markets. In 1991, he founded a hedge fund with fellow ex-Eudaemon Norman Packard. It was named Prediction Company, and the plan was to apply concepts from chaos theory to the financial world. Mixing physics and finance was to prove extremely successful, and Farmer spent eight years with the company before deciding to return to academia.

Farmer is now a professor at the University of Oxford, where he looks at the effects of introducing complexity to economics. Although there is already plenty of mathematical thinking in the world of finance, Farmer has pointed out that it is generally aimed at specific transactions. People use mathematics to decide the price of their financial products or to estimate the risk involved in certain trades. But how do all these interactions fit together? If bots influence each other's decisions, what effect could it have on the economic system as a whole? And what might happen when things go wrong?

A crisis can sometimes begin with a single sentence. At lunchtime on April 23, 2013, the following message appeared on the Associated Press's Twitter feed: "Breaking: Two Explosions in the White House and Barack Obama is injured." The news was relayed to the millions of people who follow the Associated Press on Twitter, with many of them reposting the message to their own followers.

Reporters were quick to question the authenticity of the tweet, not least because the White House was hosting a press conference at the time (which had not seen any explosions). The message indeed turned out to be a hoax, posted by hackers. The tweet was soon removed, and the Associated Press Twitter account was temporarily suspended.

Unfortunately, financial markets had already reacted to the news. Or, rather, they had overreacted. Within three minutes of the fake announcement, the S&P 500 stock index had lost $136 billion in value. Although markets soon returned to their original level, the speed—and severity—of the reaction made some financial analysts wonder whether it was really caused by human traders. Would people have really spotted an errant tweet so quickly? And would they have believed it so easily?

It wasn't the first time a stock index had ended up looking like a sharp stalactite, stretching down from the realms of sanity. One of the biggest market shocks came on May 6, 2010. When the US financial markets opened that morning, already several potential clouds were on the horizon, including the upcoming British election and ongoing financial difficulties in Greece. Yet nobody foresaw the storm that was to arrive midafternoon.

Although the Dow Jones Industrial Average had dipped a little earlier in the day, at 2:32 p.m. it started to decline sharply. By 2:42 p.m. it had lost almost 4 percent in value. The decline accelerated, and five minutes later the index was down another 5 percent. In barely twenty minutes, almost $900 billion had been wiped from the market's value. The descent triggered one of the exchange's fail-safe mechanisms, which paused trading for a few moments. This allowed prices to stabilize, and the index started to clamber back toward its original level. Even so, the drop had been staggering. So, what had happened?

Severe market disruptions can often be traced to one main trigger event. In 2013, it was the hoax Twitter announcement about the White House. Bots that scour online newsfeeds, attempting to exploit information before their competitors, would have likely picked up on this and started making trades. The story gained a curious footnote in the following year, when the Associated Press introduced automated company earnings reports. Algorithms sift through the reports and produce a couple of hundred words summarizing firms' performance in the Associated Press's traditional writing style. The change means that humans are now even more absent from the financial news process. In press offices, algorithms convert reports into prose; on trading floors, their fellow robots turn these words into trading decisions.

The 2010 Dow Jones "flash crash" was thought to be the result of a different type of trigger event: a trade rather than an announcement. At 2:32 p.m., a mutual fund had used an automated program to sell seventy-five thousand futures contracts. Instead of spreading the order over a period of time, as a series of small icebergs, the program had apparently dropped the whole thing in pretty much all at once. The previous time the fund had dealt with a trade that big, it had taken five hours to sell seventy-five thousand contracts. On this occasion, it had completed the whole transaction in barely twenty minutes.

It was undoubtedly a massive order, but it was just one order, made by a single firm. Likewise, bots that analyze Twitter feeds are relatively niche applications: the majority of banks and hedge funds do not trade in this way. Yet the reaction of these Twitter-happy algorithms led to a spike that wiped billions off the stock market. How did these seemingly isolated events lead to such turbulence?

To understand the problem, we can turn to an observation made by economist John Maynard Keynes in 1936. During the 1930s, English newspapers would often run beauty contests. They would publish a collection of girls' photos and ask readers to vote for the six they thought would be most popular overall. Keynes pointed out that shrewd readers wouldn't simply choose the girls they liked best. Instead, they would select the ones they thought everyone else would pick. And, if readers were especially sharp, they would go to the next level and try to work out which girl everyone else would expect to be the most popular.

According to Keynes, the stock market often works in much the same way. When speculating on share prices, investors are in effect trying to anticipate what everyone else will do. Prices don't necessarily rise because a company is fundamentally sound; they increase because other investors think the company is valuable. The desire to know what others are thinking means lots of second-guessing. What's more, modern markets are moving further and further away from a carefully considered newspaper contest. Information arrives fast, and so does the action. And this is where algorithms can run into trouble.

Bots are often viewed as complicated, opaque creatures. Indeed, *complex* seems to be the preferred adjective of journalists writing about trading algorithms (or any algorithm, for that matter). But in high-frequency trading, it's quite the opposite: if you want to be quick, you need to keep things simple. The more instructions you have to deal with when trading financial products, the longer things take. Rather than clogging up their bots with subtlety and nuance, creators instead limit strategies to a few lines of computer code. Doyne Farmer warns that this doesn't leave much room for reason and rationality. "As soon as you limit what you can do to ten lines of code, you're non-rational," he said. "You're not even at insect-level intelligence."

When traders react to a big event—whether a Twitter post or a major sell order—it piques the attention of the high-speed algorithms monitoring market activity. If others are selling stocks, they join in. As prices plummet, the programs follow each other's trades, driving prices further downward. The market turns into an extremely fast beauty contest, with no one wanting to pick the wrong girl. The speed of the game can lead to serious problems. After all, it's hard to work out who will move first when algorithms are faster than the eye can see. "You don't have much time to think," Farmer said. "It creates a big danger of over-reaction and herding."

Some traders have reported that mini flash crashes happen frequently. These shocks are not severe enough to grab headlines, but they are still there to be found by anyone who looks hard enough. A share price might drop in a fraction of a second, or trading activity will suddenly increase a hundredfold. In fact, there might be several such crashes every day. When researchers at the University of Miami looked at stock market data between 2006 and 2011, they found thousands of "ultrafast extreme events" in which a stock crashed or spiked in value—and recovered again—in less than a second. According to Neil Johnson, who led the research, these events are a world away from the kind of situations covered by traditional financial theories. "Humans are unable to participate in real time," he said, "and instead, an ultrafast ecology of robots rises up to take control."

WHEN PEOPLE TALK ABOUT chaos theory, they often focus on the physics side of things. They might mention Edward Lorenz and his work on forecasting and the butterfly effect: the unpredictability of the weather, and the tornado caused by the flap of an insect's wings. Or they might recall the story of the Eudaemons and roulette prediction, and how the trajectory of a billiard ball can be sensitive to initial conditions. Yet chaos theory has reached beyond the physical sciences. While the Eudaemons were preparing to take their roulette strategy to Las Vegas, on the other side of the United States ecologist Robert May was working on an idea that would fundamentally change how we think about biological systems.

Princeton University is a world away from the glittering high-rises of Las Vegas. The campus is a maze of neo-Gothic halls and sun-dappled quads; squirrels dash through ivy-clad archways, while students' distinctive orange and black scarves billow in the New Jersey wind. Look carefully and there are also traces of famous past residents. There's an "Einstein Drive," which loops in front of the nearby Institute of Advanced Study. For a while there was also a "Von Neumann corner," named after all the car accidents the mathematician reportedly had there. The story goes that von Neumann had a particularly ambitious excuse for one of his collisions. "I was proceeding down the road," he said. "The trees on the right were passing me in orderly fashion at sixty miles per hour. Suddenly one of them stepped in my path."

During the 1970s, May was a professor of zoology at the university. He spent much of his time studying animal communities. He was particularly interested in how animal numbers changed over time. To examine how different factors influenced ecological systems, he constructed some simple mathematical models of population growth.

From a mathematical point of view, the simplest type of population is one that reproduces in discrete bursts. Take insects: many species in temperate regions breed once per season. Ecologists can explore the behavior of hypothetical insect populations using an equation called "the logistic map." The concept was first proposed in 1838 by statistician Pierre Verhulst, who was investigating potential limits to population. To calculate the population density in a particular year using the logistic map, we multiply three factors together: the population growth rate, the density in the previous year, and the amount of space—and hence resources—still available. Mathematically, this takes the form:

Density in next year = Growth rate × Current density × (1–Current density)

The logistic map is built on a simple set of assumptions, and when the growth rate is small it churns out a simple result. Over a few seasons, the population settles down to equilibrium, with the population density remaining the same from one year to the next.

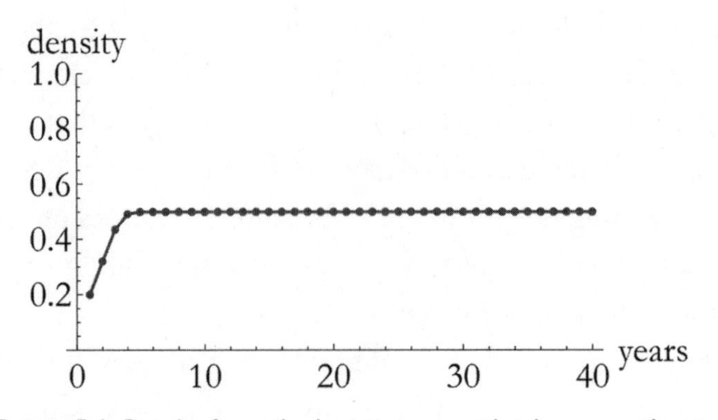

FIGURE 5.1. Results from the logistic map with a low growth rate.

The situation changes as the growth rate increases. Eventually, the population density starts to oscillate. In one year, lots of insects are hatched, which reduces available resources; next year, fewer insects survive, which makes spaces for more creatures the following year, and so on. If we sketch out how the population changes over time, we get the picture shown in Figure 5.2.

When the growth rate gets even larger, something strange happens. Rather than settle down to a fixed value, or switch between two values in a predictable way, the population density begins to vary wildly.

Remember that there is no randomness in the model, no chance events. The animal density depends on a simple one-line equation.

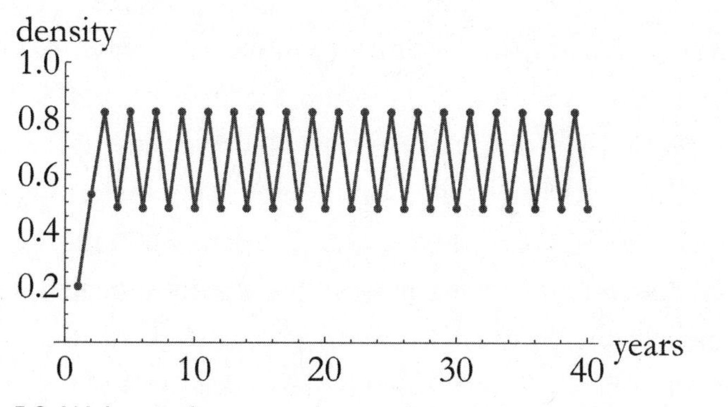

FIGURE 5.2. With a medium growth rate, the population density oscillates.

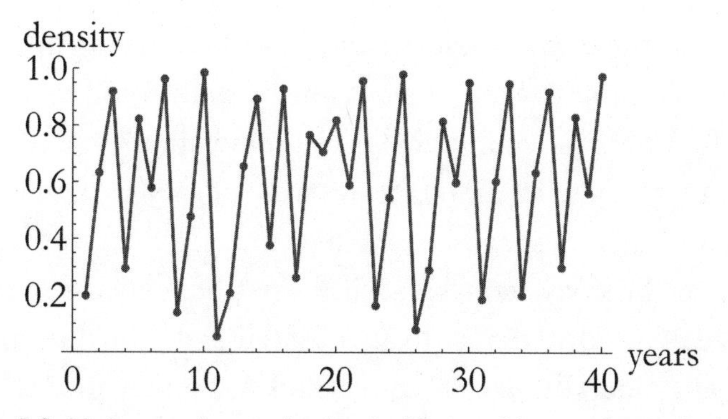

FIGURE 5.3. High growth rates lead to highly variable population dynamics.

And yet the result is a bumpy, noisy set of values, which do not appear to follow a straightforward pattern.

May found that chaos theory could explain what was going on. The fluctuations in density were the result of the population being sensitive to initial conditions. Just as Poincaré had found for roulette, a small change in the initial setup had a big effect on what happens further down the line. Despite the population following a straightforward biological process, it was not feasible to predict how it would behave far into the future.

We might expect roulette to produce unexpected outcomes, but ecologists were stunned to find that something as simple as the logistic map could generate such complex patterns. May warned that the result could have some troubling consequences in other fields, too. From politics to economics, people needed to be aware that simple systems do not necessarily behave in simple ways.

As well as studying single populations, May thought about ecosystems as wholes. For example, what happens when more and more creatures join an environment, generating a complicated web of interactions? In the early 1970s, many ecologists would have said the answer was a positive one. They believed that complexity was generally a good thing in nature; the more diversity there was in an ecosystem, the more robust it would be in the face of a sudden shock.

That was the dogma, at least, and May was not convinced it was correct. To examine whether a complex system could really be stable, he looked at a hypothetical ecosystem with a large number of interacting species. The interactions were chosen at random: some were beneficial to a species, some harmful. He then measured the stability of the ecosystem by seeing what happened when it was disrupted. Would it return to its original state, or do something completely different, like collapse? This was one of the advantages of working with a theoretical model: he could test stability without disrupting the real ecosystem.

May found that the larger the ecosystem, the less stable it would be. In fact, as the number of species grew very large, the probability of the ecosystem surviving shrank to zero. Increasing the level of complexity had a similarly harmful effect. When the ecosystem was more connected, with a higher chance of any two given species interacting with each other, it was less stable. The model suggested that the existence of large, complex ecosystems was unlikely, if not impossible.

Of course, there are plenty of examples of complex yet seemingly robust ecosystems in nature. Rainforests and coral reefs have vast numbers of different species, yet they haven't all collapsed. According to ecologist Andrew Dobson, the situation is the biological equivalent of a joke made in the early days of the European currency union. Although the euro worked in practice, observers said, it was not clear why it worked in theory.

To explain the difference between theory and reality, May suggested that nature had to resort to "devious strategies" to maintain stability. Researchers have since put forward all sorts of intricate strategies in an attempt to drag the theory closer to nature. Yet, according to Stefano Allesina and Si Tang, two ecologists at the University of Chicago, this might not be necessary. In 2013, they proposed a possible explanation for the discrepancy between May's model and real ecosystems.

Whereas May had assumed random interactions between different species—some positive, some negative—Allesina and Tang focused on three specific relationships that are common in nature. The first of these was a predator-prey interaction, with one species eating another; obviously, the predator will gain from this relationship, and the prey will lose out. As well as predation, Allesina and Tang also included cooperation, where both parties benefit from the relationship, and competition, with both species suffering negative effects.

Next, the researchers looked at whether each relationship stabilized the overall system or not. They found that excessive levels of competitive and cooperative relationships were destabilizing, whereas predator-prey relationships had a stabilizing effect on the system. In other words, a large ecosystem could be robust to disruption as long as it had a series of predator-prey interactions at its core.

So, what does all this mean for betting and financial markets? Much like ecosystems, markets are now inhabited by several different bot species. Each has a different objective and specific strengths and weaknesses. There are bots out hunting for arbitrage opportunities; they are trying to react to new information first, be it an important event or an incorrect price. Then there are the "market makers," offering to accept trades or bets on both sides and pocket the difference. These bots are essentially bookmakers, making their money by anticipating where the action will be. They buy low and sell high, with the aim of balancing their books. There are also bots trying to hide large transactions by sneaking smaller trades into the market. And there are predator bots watching for these large trades, hoping to spot a big transaction and take advantage of the subsequent shift in the market.

During the flash crash on May 6, 2010, there were over fifteen thousand different accounts trading the futures contracts involved in the crisis. In a subsequent report, the Securities and Exchange Commission (SEC) divided the trading accounts into several different categories, depending on their role and strategy. Although there has been much debate about precisely what happened that afternoon, if the crash was indeed triggered by a single event—as the SEC report suggested—the havoc that followed was not the result of one algorithm. Chances are it came from the interaction between lots of different trading programs, with each one reacting to the situation in its own way.

Some interactions had particularly damaging effects during the flash crash. In the middle of the crisis, at 2:45 p.m., there was a drought of buyers for futures contracts. High-frequency algorithms therefore traded among themselves, swapping over twenty-seven thousand futures in the space of fourteen seconds. Normality only resumed after the exchange deliberately paused the market for a few seconds, halting the runaway drop in price.

Rather than treating betting or financial markets as a set of static economic rules, it makes sense to view them as an ecosystem. Some traders are predators, feeding off weaker prey. Others are competitors, fighting over the same strategy and both losing out. Many of the ideas and warnings from ecology can therefore apply to markets. Simplicity does not mean predictability, for example. Even if algorithms follow simple rules, they won't necessarily behave in simple ways. Markets also involve webs of interactions—some strong, some brittle—which means that having lots of different bots in the same place does not necessarily help matters. Just as May showed, making an ecosystem more complex doesn't necessarily make it more stable.

Unfortunately, increased complexity is inevitable when there are lots of people looking for profitable strategies. Whether in betting or finance, ideas are less lucrative once others notice what is going on. As exploitable situations become widely known, the market gets more efficient and the advantage disappears. Strategies therefore have to evolve as existing approaches become redundant.

Doyne Farmer has pointed out that the process of evolution can be broken down into several stages. To come up with a good strategy, you first need to spot a situation that can be exploited. Next, you need to get ahold of enough data to test whether your strategy works. Just as gamblers need plenty of data to rate horses or sports teams, traders need enough information to be sure that the advantage is really there, and not a random anomaly. At Prediction Company, this process was entirely algorithm-driven. The trading strategies were what Farmer called "evolving automata," with the decision-making process mutating as the computers accumulated new experience.

The shelf life of a trading strategy depends on how easy it is to complete each evolutionary stage. Farmer has suggested that it can often take years for markets to become efficient and strategies to become useless. Of course, the bigger the inefficiency is, the easier it is to spot and exploit. Because computer-based strategies tend to be highly lucrative at first, copycats are more likely to appear. Algorithmic approaches therefore have to evolve faster than other types of strategy. "There's going to be an ongoing saga of one-upmanship," Farmer said.

RECENT YEARS HAVE SEEN a huge growth in the number of algorithms scouring financial markets and betting exchanges. It is the latest connection between two industries that have a history of shared ideas, from probability theory to arbitrage. But the distinction between finance and gambling is blurring more than ever before.

Several betting websites now allow people to bet on financial markets. As with other types of online betting, these transactions constitute gambling and hence are exempt from tax in many European countries (at least for the customer; there is still a tax burden on the bookmaker). One of the most popular types of financial wager is spread betting. In 2013, around a hundred thousand people in Britain placed bets in this way.

In a traditional bet, the stake and potential payoff are fixed. You might bet on a certain team winning or on a share price rising. If the outcome goes your way, you get the payoff. If not, you lose your stake. Spread betting is slightly different. Your profit depends not just on the outcome but also on the size of the outcome. Let's say a share is currently priced at $50, and you think it will increase in value in the next week. A spread betting company might offer you a spread bet at $1 per point over $51 (the difference between the current price and the offered number is the "spread," and how the bookmaker makes its money). For every dollar the price rises above $51, you will get $1, and for every dollar it drops below, you will lose $1. In terms of payoff, it's not that different from simply buying the share and then selling it a week later. You'll make pretty much the same amount of profit (or loss) on both the bet and the financial transaction.

But there is a crucial difference. If you make a profitable stock trade in the United Kingdom, you have to pay stamp duty and capital gains tax. If you place a spread bet, you don't. Things are different in other countries. In Australia, profits from spread betting are classed as income and are therefore subject to tax.

Deciding how to regulate transactions is a challenge in both gambling and finance. When dealing with an intricate trading ecosystem, however, it is not always clear what effects regulation will have. In 2006, the US Federal Reserve and the National Academy of Sciences brought together financiers and scientists to debate "systemic risk" in finance. The idea was to consider the stability of the financial system as a whole rather than just the behavior of individual components.

During the meeting, Vincent Reinhart, an economist at the Federal Reserve, pointed out that a single action could have multiple potential outcomes. The question, of course, is which one will prevail. The result won't depend on only what regulators do. It could also depend on how the policy is communicated and how the market reacts to news. This is where economic approaches borrowed from the physical sciences can come up short. Physicists study interactions that follow known rules; they don't generally have to deal with human behavior. "The odds on a hundred-year storm do not change because people think it has become more likely," Reinhart said.

Ecologist Simon Levin, who also attended the meeting, elaborated on the unpredictability of behavior. He noted that economic interventions—like the ones available to the Federal Reserve—aim to change individual behavior in the hope of improving the system as a whole. Although certain measures can change what individuals do, it is very difficult to stop panic spreading through a market.

Yet the spread of information is only going to get faster. News no longer has to be read and processed by humans. Bots are absorbing news automatically and handing it to programs that make trading decisions. Individual algorithms react to what others do, with decisions made on the sort of timescales that humans can never fully supervise. This can lead to dramatic, unexpected behavior. Such problems often come from the fact that high-frequency algorithms are designed to be simple and fast. The bots are rarely complex or clever: the aim is to exploit an advantage before anyone else gets there. Creating successful artificial gamblers is not always a matter of being first, however. As we shall discover, sometimes it pays to be smart.

6

LIFE CONSISTS OF BLUFFING

IN SUMMER 2010, POKER WEBSITES LAUNCHED A CRACKDOWN ON robot players. By pretending to be people, these bots had been winning tens of thousands of dollars. Naturally, their human opponents weren't too happy. In retaliation, website owners shut down any accounts that were apparently run by software. One company handed almost $60,000 back to players after discovering that bots had been winning on their tables.

It wasn't long before computer programs again surfaced in online poker games. In February 2013, Swedish police started investigating poker bots that had been operating on a state-owned poker website. It turned out that these bots had made the equivalent of over half a million dollars. It wasn't just the size of the haul that worried poker companies; it was how the money was made. Rather than taking money from weaker players in low-stakes games, the bots had been winning on high-stakes tables. Until these sophisticated computer players were discovered, few people in the industry had realized that bots were capable of playing so well.

Yet poker algorithms have not always been so successful. When bots first became popular in the early 2000s, they were easily beaten. So, what has changed in recent years? To understand why bots are getting better at poker, we must first look at how humans play games.

WHEN THE US CONGRESS put forward a bill in 1969 suggesting that cigarette advertisements be banned from television, people expected American tobacco companies to be furious. After all, this was an industry that had spent over $300 million promoting their products the previous year. With that much at stake, a clampdown would surely trigger the powerful weapons of the tobacco lobby. They would hire lawyers, challenge members of Congress, fight antismoking campaigners. The vote was scheduled to take place in December 1970, which gave the firms eighteen months to make their move. So, what did they choose to do? Pretty much nothing.

Far from hurting tobacco companies' profits, the ban actually worked in the companies' favor. For years, the firms had been trapped in an absurd game. Television advertising had little effect on whether people smoked, which in theory made it a waste of money. If the firms had all got together and stopped their promotions, profits would almost certainly have increased. However, ads did have an impact on which brand people smoked. So, if all the firms stopped their publicity, and one of them started advertising again, that company would steal customers from all the others.

Whatever their competitors did, it was always best for a firm to advertise. By doing so, it would either take market share from companies that didn't promote their products or avoid losing customers to firms that did. Although everyone would save money by cooperating, each individual firm would always benefit by advertising.

Which meant all the companies inevitably ended up in the same position, putting out advertisements to hinder the other firms. Economists refer to such a situation—where each person is making the best decision possible given the choices made by others—as a "Nash equilibrium." Spending would rise further and further until this costly game stopped. Or somebody forced it to stop.

Congress finally banned tobacco ads from television in January 1971. One year later, the total spent on cigarette advertising had fallen by over 25 percent. Yet tobacco revenues held steady. Thanks to the government, the equilibrium had been broken.

JOHN NASH PUBLISHED HIS first papers on game theory while he was a PhD student at Princeton. He'd arrived at the university in 1948, after being awarded a scholarship on the strength of his undergraduate tutor's reference, a two-sentence letter that read, "Mr. Nash is nineteen years old and is graduating from Carnegie Tech in June. He is a mathematical genius."

During the next two years, Nash worked on a version of the "prisoner's dilemma." This hypothetical problem involves two suspects caught at the scene of a crime. Each is placed in a separate cell and must choose whether to remain silent or testify against the other person. If they both keep quiet, both receive one-year sentences. If one remains silent and the other talks, the quiet prisoner gets three years and the one who blames him is released. If both talk, both are sent down for two years.

Overall, it would be best if both prisoners kept their mouths shut and took the one-year sentence. However, if you are a prisoner stuck alone in a cell, unable to tell what your accomplice is going to do, it is always better to talk: if your partner stays silent, you get off; if your partner talks, you receive two years rather than three. The Nash equilibrium for the prisoner's dilemma game therefore has both players talking. Although they will end up suffering two years in prison rather than one, neither will gain anything if one alone changes strategy. Substitute talking and silence for advertising and cutting promotions, and it is the same problem the advertising firms faced.

Nash received his PhD in 1950, for a twenty-seven-page thesis describing how his equilibrium can sometimes thwart seemingly beneficial outcomes. But Nash wasn't the first person to take a mathematical hammer to the problem of competitive games. History has given that accolade to John von Neumann. Although later known for his time at Los Alamos and Princeton, in 1926 von Neumann was a young lecturer at the University of Berlin. In fact, he was the youngest in its history. Despite his prodigious academic record, however, there were still some things he wasn't very good at. One of them was poker.

Poker might seem like the ideal game for a mathematician. At first glance, it's just a matter of probabilities: the probability you receive a good hand; the probability your opponent gets a better one. But anyone who has played poker using only probability knows that things are not so simple. "Real life consists of bluffing," von Neumann noted, "of little tactics of deception, of asking yourself what is the other man going to think I mean to do." If he was to grasp poker, he would need to find a way to account for his opponent's strategy.

Von Neumann started by looking at poker at its most basic, where it is a game between two players. To simplify matters further, he assumed that each player was dealt a single card, showing a number somewhere between 0 and 1. After both players put in a dollar to start, the first player—who we'll call Alice—has three options: fold, and therefore lose one dollar; check (equivalent to betting nothing); or bet one dollar. Her opponent then decides whether to fold, and forfeit the money, or match the bet, in which case the winner depends on whose card has the highest number.

Obviously, it is pointless for Alice to fold at the start, but should she check or bet? Von Neumann looked at all possible eventualities and worked out the expected profit from each strategy. He found that she should bet if her card shows a very low or very high number, and she should check otherwise. In other words, she should bluff only with her worst hand. This might seem counterintuitive, but it follows logic familiar to all good poker players. If her card shows an average-to-low number, Alice has two options: bluff or check. With a terrible card, Alice has no hope of winning unless her opponent folds. She should therefore bluff. Middling cards are trickier. Bluffing won't persuade someone with a decent card to fold, and it's not worth Alice betting on the off chance that her mediocre card will come out on top in a showdown. So, the best option is to check and hope for the best.

In 1944, von Neumann and economist Oskar Morgenstern published their insights in a book titled *Theory of Games and Economic Behavior.* Although their version of poker was much simpler than the real thing, the pair had cracked a problem that had long bothered players, namely, whether bluffing was really a necessary part of the game. Thanks to von Neumann and Morgenstern, there was now mathematical proof that it was.

Despite his fondness for Berlin's nightlife, von Neumann didn't use game theory when he visited casinos. He saw poker mainly as an intellectual challenge and eventually moved on to other problems. It would be several decades before players worked out how to use von Neumann's ideas to win for real.

BINION'S GAMBLING HALL IS part of the old Las Vegas. Away from the Strip's shows and fountains, it lies in the thumping downtown heart of the city. While most hotels were built with theaters and concert halls as well as casinos, Binion's was designed for gambling from the start. When it opened in 1951, betting limits were much higher than at other venues, and in the entrance a giant upturned horseshoe straddled a box displaying a million dollars in cash. Binion's was also the first casino to give free drinks to all gamblers to keep them (and their money) at the tables. So, it was only natural that when the first World Series of Poker took place in 1970, it was held at Binion's.

Over the following decades, players gathered at Binion's each year to pit their wits—and luck—against each other. Some years were especially tense. Early in the 1982 competition, Jack Straus stumbled onto a losing streak that left him with a single chip. Fighting back, he managed to win enough hands to stay in the game, eventually going on to win the whole tournament. The story goes that when Straus was later asked what a poker player needs for victory, his reply was "a chip and a chair."

On May 18, 2000, the thirty-first World Series reached its finale. Two men were left in the competition. On one side of the table was T. J. Cloutier, a poker veteran from Texas. Opposite him sat Chris Ferguson, a long-haired Californian with a penchant for cowboy hats and sunglasses. Ferguson had started the game with far more chips than Cloutier, but his lead was shrinking with each hand that was dealt.

With the players almost even, the dealer handed out yet another set of cards. They were playing Texas hold'em poker, which meant that Ferguson and Cloutier first received two personal "pocket" cards. After looking at his hand—the ninety-third of the day—Cloutier opened with a bet of almost $200,000. Sensing a chance to retake the advantage, Ferguson raised him half a million dollars. But Cloutier was confident, too. So confident, in fact, that he responded by pushing all his chips into the center of the table. Ferguson looked at his cards again. Did Cloutier really have the better hand? After pondering his options for several moments, Ferguson decided to match Cloutier's bet of almost $2.5 million.

Once two initial pocket cards have been dealt in Texas hold'em, there are up to three additional rounds of betting. The first of these is known as the "flop." Three more cards are dealt, this time placed face up on the table. If betting continues, another card—the "turn"—is revealed. Another round of betting means that the game reaches the "river," where a fifth card is shown. The winner is the player who has the best five-card hand when the two pocket cards are combined with the five communal cards.

Because Cloutier and Ferguson had both gone all in at the start, there would be no additional betting. Instead, they would have to show their pocket cards and watch as the dealer turned over each of the five additional cards. When the players showed their hands, the crowd surrounding the table knew Ferguson was in trouble. Cloutier had an ace and a queen; Ferguson had only an ace and a nine. First, the dealer turned over the flop cards: a king, a two, and a four. Cloutier still had the better hand. Next came the turn, and another king. The game would therefore be settled on the river. As the final card was revealed, Ferguson leapt from his seat. It was a nine. He'd won the game, and the tournament. "You didn't think it would be that tough to beat me, did you?" Cloutier asked Ferguson after he'd netted the $1.5 million prize money. "Yes," Ferguson replied, "I did."

UNTIL CHRIS FERGUSON'S TRIUMPHANT performance in Las Vegas, no poker player had won more than $1 million in tournament prizes. But unlike many competitors, Ferguson's extraordinary success did not rely solely on intuition or instinct. When he played in the World Series, he was using game theory.

The year before he beat Cloutier, Ferguson had completed a doctorate in computer science at UCLA. During that time, he worked as a consultant for the California State Lottery, picking apart existing games and coming up with new ones. His family members have mathematical backgrounds, too: both parents have PhDs in the subject and his father, Thomas, is a professor of mathematics at UCLA.

While studying for his doctorate, Chris Ferguson would compete for play money in some of the early Internet chat rooms. He saw poker as a challenge, and it was one he was rather good at. The chat room games didn't lead to any profit, but they did give Ferguson access to large amounts of data. Combined with improvements in computing power, this enabled him to study vast numbers of different hands, evaluating how much to bet and when to bluff.

Like von Neumann, Ferguson soon realized that poker was too complicated to study properly without making a few simplifications. Building on von Neumann's ideas, Ferguson decided to look at what happens when two players have more options. Of course, he would have more than one opponent at the start of a real poker game, but it was still useful to analyze the simple two-player scenario. Players may fold as the betting rounds progress, so by the time the endgame arrives, there are often only a couple of players left.

Yet there are still a number of things the two players might do at this point. The first player, Alice, had three simple choices in von Neumann's game—bet one dollar, check, or fold—but in a real game she might do something else, like change her bet. And the second player might not respond by matching the bet or folding. The second player might be confident like Cloutier was and raise the betting.

As more options creep into the game, picking the best one becomes more complicated. In a simple setup, von Neumann showed that players should employ "pure strategies," in which they follow fixed rules such as "if this happens, always do A" and "if that happens, always do B." But pure strategies are not always a good approach to use. Take a game of rock-paper-scissors. Picking the same option every time is admirably consistent, but the strategy is easy to beat if your opponent works out what you're doing. A better idea is to use a "mixed strategy." Rather than always going with the same approach, you should switch between one of the pure strategies—rock, paper, or scissors—with a certain probability. Ideally, you will play each of the three options in a balance that makes it impossible for your opponent to guess what you're going to do. For rock-paper-scissors, the optimal strategy against a new opponent is to choose randomly, playing each option one-third of the time.

Mixed strategies also make an appearance in poker. Analysis of the endgame suggests that you should balance the number of times you are honest and the number of times you bluff so that your opponent is indifferent to calling or folding. Like rock-paper-scissors, you don't want the other person to work out what you are likely to do. "You always want to make your opponents' decisions as difficult as possible," Ferguson said.

Sifting through the data from the chat room games, Ferguson spotted other areas for improvement. When experienced players had good hands, they would raise heavily to encourage their opponents to fold. This removed the risk of a weak hand turning into a winning hand when the communal cards were revealed. But Ferguson's research showed that the raises were too high: sometimes it was worth betting less and allowing people to remain in the game. As well as winning more money with strong hands, it meant that if a hand did lose, it wouldn't lose as much.

Through his research, Ferguson discovered that finding a successful approach to poker doesn't necessarily mean chasing profits at all costs. As he once told *The New Yorker*, the optimal strategy isn't a case of "How do I win the most?" but one of "How do I lose the least?" Novice players usually confuse the two and don't fold often enough as a result. True, it's impossible to win anything by folding, but sitting out a hand allows players to avoid costly betting rounds. Collecting together his results into detailed tables, Ferguson memorized the strategies—including when to bluff, when to bet, how much to raise—and started playing for real money. He entered his first World Series in 1995; five years later he was champion.

Ferguson has always been fond of picking up new skills. He once taught himself to throw a playing card so fast from a distance of ten feet that it could slice a carrot in two. In 2006, he decided to take on a new challenge. Starting with nothing, he would work his way up to $10,000. His aim was to show the importance of bankroll management in poker. Just as the Kelly criterion helped gamblers adjust their bet size in blackjack and sports betting, Ferguson knew it was essential to adjust his playing style to balance profit and risk.

Because he was starting with zero dollars, Ferguson's first task was to get ahold of some cash. Fortunately, some poker websites ran daily "freeroll tournaments." Hundreds of players could enter for free, with the top dozen or so receiving cash prizes. It's not often that a big-name player enters a freeroll tournament, let alone takes it seriously. When other online players found out who they were playing against, most thought it was a joke. Why was a world champion like Chris Ferguson plying his trade on the free tables?

After a few attempts, Ferguson eventually netted some all-important cash. "I remember winning my first $2 a couple of weeks into the challenge," he later wrote, "and I strategized for three days, deliberating over what game to play with it." He settled on the lowest-stakes game possible, but within one round he'd lost it all. Finding himself back to zero, he returned to the freeroll tournaments and started over again. It was clear that he would have to be extremely disciplined if he was going to reach his target.

Playing around ten hours per week, it took Ferguson nine months to get to $100 (he'd expected it to take around six). He kept going, sticking to a strict set of rules. For instance, he would only ever risk 5 percent of his bankroll in a particular game. It meant that if he lost a few rounds, he would have to go back to lower-stakes tables. Psychologically, he found it difficult to drop down a level. Ferguson was used to the excitement of high-stakes games and the profits they brought. After moving down, he would lose focus and struggle to keep to his rules. Rather than take more risks, he stepped back; it was pointless playing the game until he'd regained his concentration. The self-restraint paid off. After another nine months of careful play, Ferguson finally reached his $10,000 total.

The bankroll challenge, along with his earlier World Series victory, cemented Ferguson's reputation as a virtuoso of poker theory. Much of his success came from working on optimal strategies, but do such strategies always exist in games like poker? The question was actually one of the first that von Neumann asked when he started working on two-player games at the University of Berlin. As well as laying the foundations for the entire field, the answer would go on to cause a bitter dispute about who was the true inventor of game theory.

GAMES LIKE POKER ARE "zero-sum," with winning players' profits equal to other players' losses. When two players are involved, this means one person is always trying to minimize the opponent's payoff—a quantity the opponent will be trying to maximize. Von Neumann called it the "minimax" problem and wanted to prove that both players could find an optimal strategy in this tug-of-war. To do this, he needed to show that each player could always find a way to minimize the maximum amount that could potentially be lost, regardless of what their opponent did.

One of the most prominent examples of a zero-sum game with two players is a soccer penalty. This ends either in a goal, with the kicker winning and the goalkeeper losing, or a miss, in which case the payoffs are reversed. Keepers have very little time to react after a penalty is taken, so generally make their decision about which way to dive before the kicker strikes the ball.

Because players are either right- or left-footed, choosing the right- or left-hand side of the goal can alter their chances of scoring. When Ignacio Palacios-Heurta, an economist at Brown University, looked at all the penalties taken in European leagues between 1995 and 2000, he found that the probability of a goal varies depending on whether the kicker chooses the "natural" half of the goal. (For a right-footed player, this would be the left-hand side of the goal; for a left-footed kicker, it would be the right side.)

The penalty data showed that if the kicker picked the natural side and the keeper chose the correct direction, the kicker scored about 70 percent of the time; if the keeper got it wrong, around 90 percent of shots went in. In contrast, kickers who went for the nonnatural side scored 60 percent of shots if the keeper picked correctly and 95 percent if they didn't. These probabilities are summarized in Table 6.1.

If kickers want to minimize their maximum loss, they should therefore pick the natural side: even if the goalkeeper gets the correct direction, the player has at least a 70 percent chance of scoring. In contrast, the keeper should dive to the kicker's nonnatural side. At worst it will result in the player scoring 90 percent of the time rather than 95 percent.

If these strategies were optimal, the worse-case probabilities for kicker and keeper would be equal. This is because a penalty shootout is zero-sum: each person is trying to minimize the potential loss, which means if each plays the perfect strategy, it should minimize the maximum payoff for the opponent. Yet this is clearly not the

TABLE 6.1. The Probability of Scoring a Penalty Depends on Which Side the Kicker and Goalkeeper Choose

		Keeper	
		Natural	Non-natural
Kicker	Natural	70%	90%
	Non-natural	95%	60%

case, because the worst outcome for the player results in scoring 70 percent of the kicks, whereas the worst result for the goalkeeper leads to letting in 90 percent of shots.

The fact that the values are not equal implies that each person can adjust tactics to improve the chances of success. As in rock-paper-scissors, switching between options might be better than relying on a simple pure strategy. For example, if the kicker always chooses the natural side, the goalkeeper should occasionally pick that option, too, which would bring the 90 percent worst-case scenario down closer to 70 percent. In response, the kicker could counter this tactic by also opting for a mixed strategy.

When Palacios-Heurta calculated the best approach for the kicker and goalkeeper, he found that both should choose the natural half of the goal with 60 percent probability, and the other side the rest of the time. Like effective bluffing in poker, this would have the effect of making the other person indifferent to what is going to happen: opponents would be unable to boost their chances by changing their strategy. Both the goalkeeper and the kicker would therefore successfully limit their loss as well as minimize the other person's gain. Remarkably, the recommended 60 percent value is within a few percent of the real proportion of times players choose each side, suggesting that—whether aware of it or not—kickers and goalkeepers have already figured out the optimal strategy for penalties.

VON NEUMANN COMPLETED HIS solution to the minimax problem in 1928, publishing the work in an article titled "Theory of Parlour Games." Proving that these optimal strategies always existed was a crucial breakthrough. He later said that without the result, there would have been no point continuing his work on game theory.

The method von Neumann used to attack the minimax problem was far from simple. Lengthy and elaborate, it has been described as a mathematical "tour de force." But not everyone was impressed. Maurice Fréchet, a French mathematician, argued that the mathematics behind von Neumann's minimax work had already been in place (though von Neumann had apparently been unaware of it). By applying the techniques to game theory, he said that von Neumann had "simply entered an open door."

The approaches Fréchet was referring to were the brainchild of his colleague Émile Borel, who had developed them a few years before von Neumann. When Borel's papers were eventually published in English in the early 1950s, Fréchet wrote an introduction crediting him with the invention of game theory. Von Neumann was furious, and the pair exchanged barbed comments in the economics journal *Econometrica*.

The dispute raised two important issues about applying mathematics to real-world problems. First, it can be hard to pin down the initiator of a theory. Should credit go to the researcher who crafts the mathematical bricks or to the person who assembles them into a useful structure? Fréchet clearly thought that brick maker Borel deserved the accolades, whereas history has given the credit to von Neumann for using mathematics to construct a theory for games.

The argument also showed that major results aren't always appreciated in their original format. Despite his defense of Borel's work, Fréchet didn't think the minimax work was particularly special because mathematicians already knew about the idea, albeit in a different form. It was only when von Neumann applied the minimax concept to games that its value become apparent. As Ferguson discovered when he applied game theory to poker, sometimes an idea that seems unremarkable to scientists can prove extremely powerful when used in a different context.

While the fiery debate between von Neumann and Fréchet sparked and crackled, John Nash was busy finishing his doctorate at Princeton. By establishing the Nash equilibrium, he had managed to extend von Neumann's work, making it applicable to a wider number of situations. Whereas von Neumann had looked at zero-sum games with two players, Nash showed that optimal strategies exist even if there are multiple players and uneven payoffs. But knowing perfect strategies always exist is just the start for poker players. The next problem is working out how to find them.

MOST PEOPLE WHO HAVE a go at creating poker bots don't rummage through game theory to find optimal strategies. Instead, they often start off with rule-based approaches. For each situation that could crop up in a game, the creator puts together a series of "if this happens, do that" instructions. The behavior of a rule-based bot therefore depends on its creator's betting style and how the creator thinks a good player should act.

While earning his master's degree in 2003, computer scientist Robert Follek put together a rule-based poker program called SoarBot. He built it using a set of artificial decision-making methods known as "Soar," which had been developed by researchers at the University of Michigan. During a poker game, SoarBot acted in three phases. First, it noted the current situation, including the pocket cards it had been dealt, the values of the communal cards, and the number of players who had folded. With this information, it then ran through all its preprogrammed rules and identified all those that were relevant to the present situation.

After collecting the available options, it entered a decision phase, choosing what to do based on preferences Follek had given it. This decision-making process could be problematic. Occasionally, the set of preferences turned out to be incomplete, with SoarBot either failing to identify any suitable options or being unable to choose between two potential moves. The predetermined preferences could also be inconsistent. Because Follek had input each one individually, sometimes the program would end up containing two preferences that were contradictory. For instance, one rule might tell SoarBot to bet in a given situation while another simultaneously tried to get it to fold.

Even if more rules were added manually, the program would still hit upon an inconsistency or incompleteness from time to time. This type of problem is well known to mathematicians. One year after obtaining his PhD in 1930, Kurt Gödel published a theorem pointing out that the rules that governed arithmetic could not be both complete and consistent. His discovery shook the research community. At the time, leading mathematicians were trying to construct a robust system of rules and assumptions for the subject. They hoped this would clear up a few logical anomalies that had recently been spotted. Led by David Hilbert, who had been von Neumann's mentor in Germany, these researchers wanted to find a set of rules that was complete—so that all mathematical statements could be proved using only these rules—and consistent, with none of the rules contradicting one another. But Gödel's incompleteness theorem showed that this was impossible: whichever set of rules was specified, there would always be situations in which additional rules were needed.

Gödel's logical rigor caused problems outside academia, too. While studying for his American citizenship assessment in 1948, he told his sponsor Oskar Morgenstern that he'd spotted some inconsistencies in the US Constitution. According to Gödel, the contradictions created a legal path for someone to become a dictator. Morgenstern told him it would be unwise to bring it up in the interview.

FORTUNATELY FOR FOLLEK, THE team that originally developed the Soar technology had found a way around Gödel's problem. When a bot ran into trouble, it would teach itself an additional rule. So, if Follek's SoarBot couldn't decide what to do, it could instead pick an arbitrary option and add the choice to its set of rules. Next time the same situation popped up, it could simply search through its memory to find out what it did last time. This type of "machine learning," with the bot adding new rules as it went along, allowed it to avoid the pitfalls Gödel described.

When Follek let SoarBot compete against human and computer opponents, it became clear his program wasn't a potential champion. "It played much better than the worst human players," he said, "and much worse than the best human and software players." Actually, SoarBot played about as well as Follek did. Although he'd read up on poker strategies, his weakness as a player limited the success of his bot.

From 2004 onward, poker bots grew in popularity thanks to the arrival of cheap software that let players put together their own bot. By tweaking the settings, they could decide which rules the program should follow. With well-chosen rules, these bots could beat some opponents. But, as Follek found, the structure of rule-based bots means that they are generally only as good as their creator. And judging by bots' success rates online, most creators aren't very good at poker.

BECAUSE RULE-BASED TACTICS CAN be difficult to get right, some people have turned to game theory to improve their computer players. But it's tricky to find the optimal tactics for a game as complicated as Texas hold'em poker. Because a huge number of different possible situations could arise, it is very difficult to compute the ideal Nash equilibrium strategy. One way around the problem is to simplify things, creating an abstract version of the game. Just as stripped-down versions of poker helped von Neumann and Ferguson understand the game, making simplifications can also help find tactics that are close to the true optimal strategy.

A common approach is to collect similar poker hands into "buckets." For example, we could work out the probability a given pair of pocket cards would beat a random other hand in a showdown, and then put hands with comparable winning probabilities into the same bucket. Such approximations dramatically reduce the number of potential scenarios we have to look at.

Bucketing also crops up in other casino games. Because the aim of blackjack is to get as near to twenty-one as possible, knowing whether the next card is likely to be high or low can give a player an advantage. Card counters get ahold of this information by keeping track of the cards that have already been dealt, and hence which ones remain. But with casinos using up to six decks at once, it's impractical to memorize each individual card as it appears. Instead, counters often group cards into categories. For instance, they might split them into three buckets: high, low, and neutral. As the game progresses, they keep a count of the type of cards they have already seen. When a high card is dealt, they add one to the count; when a low card arrives, they subtract one.

In blackjack, bucketing provides only an estimate of the true count: the fewer buckets a player uses, the less accurate the count will be. Likewise, bucketing won't give poker players a perfect strategy. Rather, it leads to what are known as "near-equilibrium strategies," some of which are closer to the true optimum than others. Just as penalty takers can improve their odds by deviating from a simple pure strategy, these not-quite-perfect poker strategies can be categorized by how much players would gain by altering their tactics.

Even with the inclusion of bucketing, we still need a way to work out a near-equilibrium strategy for poker. One way to do this is to use a technique known as "regret minimization." First, we create a virtual player and give it a random initial strategy. So, it might start off by folding half of the time in a given situation, betting the other half, and never checking. Then we simulate lots and lots of games and allow the player to update its strategy based on how much it regrets its choices. For instance, if the opponent folds prematurely, the player might regret betting big. Over time, the player will work to minimize the amount of regret it has, and in the process approach the optimal strategy.

Minimizing regret means asking, "How would I have felt if I'd done things differently?" It turns out that the ability to answer this question can be critical when playing games of chance. In 2000, researchers at the University of Iowa reported that people who had injuries to parts of the brain that are related to regret—such as the orbitofrontal cortex—performed very differently in betting games compared to those without brain damage. It wasn't because the injured players failed to remember previous bad decisions. In many cases, patients with orbitofrontal damage still had a good working memory: when asked to sort a series of cards, or match up different symbols, they had few problems. The difficulties came when they had to deal with uncertainty and use their past experiences to weigh the risks involved. The researchers found that when the feeling of regret was missing from patients' decision-making process, they struggled to master games involving an element of risk. Rather than simply looking forward to try to maximize payoffs, it appears that it is sometimes necessary to look back at what might have happened and use hindsight to refine a strategy. This contrasts with much economic theory, in which the focus is often on expected gains, with people trying to maximize future payoffs.

Regret minimization is becoming a powerful tool for artificial players. By repeatedly playing games and reevaluating past decisions, bots can construct near-equilibrium strategies for poker. The resulting strategies are far more successful than simple rule-based methods. Yet such approaches still rely on making estimates, which means that against a perfect poker bot, a near-equilibrium strategy will struggle. But how easy is it to make a perfect bot for a complex game?

GAME THEORY WORKS BEST in straightforward games in which all information is known. Tic-tac-toe is a good example: after a few games, most people work out the Nash equilibrium. This is because there aren't many ways in which a game can progress: if a player gets three in a row, the game is over; players must take it in turns; and it doesn't matter which way the board is oriented. So, although there are 3^9 ways to place *X*'s and *O*'s on a three-by-three board, only a hundred or so of these 19,683 combinations are actually relevant.

Because tic-tac-toe is so simple, it's fairly easy to work out the perfect way to react to an opponent's move. And once both players know the ideal strategy, the game will always result in a draw. Checkers, however, is far from simple. Even the best players have failed to find the perfect strategy. But if anyone could have spotted it, it would have been Marion Tinsley.

A mathematics professor from Florida, Tinsley had a reputation for being unbeatable. He won his first world championship in 1955, holding it for four years before choosing to retire, citing a lack of decent competition. Upon his return to the championships in 1975, he immediately regained his title, thrashing all opposition. Fourteen years later, however, Tinsley's interest in the game had again started to wane. Then he heard about a piece of software being developed at the University of Alberta in Canada.

Jonathan Schaeffer is now Dean of Science at the university, but back in 1989 he was a young professor in the department of computer science. He'd become interested in checkers after spending time looking at chess programs. Like chess, checkers is played on an eight-by-eight board. Pieces move forward diagonally and capture opposition pieces by leapfrogging. Upon reaching the other side of the board, they become kings and can move backward and forward. The simplicity of its rules makes checkers interesting to game theorists because it is relatively easy to understand, and players can predict the consequences of a move in detail. Perhaps a computer could even be trained to win?

Schaeffer decided to name the fledgling checkers project "Chinook" after the warm winds that occasionally sweep over the Canadian prairies. The name was a pun, inspired by the fact that the English call the game "draughts." Helped by a team of fellow computer scientists and checkers enthusiasts, Schaeffer quickly got to work on the first challenge: how to deal with the complexity of the game. There are around 10^{20} possible positions in checkers. That's 10 followed by twenty zeros: if you collected the sand from all of the world's beaches, you'd end up with about that number of grains.

To navigate this enormous selection of possibilities, the team got Chinook to follow a minimax approach, hunting down strategies that would be least costly. At each point in the game, there were a certain number of moves Chinook could make. Each of these branched out into another set of options, depending on what its opponent did. As the game progressed, Chinook "pruned" this decision tree, removing weak branches that were likely to lose it the game and examining stronger, potentially winning, branches in detail.

Chinook also had a few tricks lined up especially for human opponents. When it spotted strategies that would eventually lead to a draw against a perfect computer opponent, it didn't necessarily ignore them. If the draw lay at the end of a long, tangled branch of options, there was a chance a human might make a mistake somewhere along the way. Unlike many game-playing programs, Chinook would often pick these antihuman strategies over an option that was actually better according to game theory.

Chinook played its first tournament in 1990, coming in second in the US National Checker Championship. This should have meant it qualified for the World Championships, but the American Checkers Federation and the English Draughts Association did not want a computer to compete. Fortunately, Tinsley didn't share their view. After a handful of unofficial games in 1990, he decided that he liked Chinook's aggressive playing style. Whereas human players would try to force a draw against him, Chinook took risks. Determined to play the computer in a tournament, Tinsley resigned his championship title. Reluctantly, the authorities decided to allow the computer match, and in 1992 Chinook played Tinsley in a "Man-Machine World Championship." Out of 39 games, Tinsley won 4 to Chinook's 2, with 33 draws.

Despite holding their own against Tinsley, Schaeffer and his team wanted to do one better. They wanted to make Chinook unbeatable. Chinook depended on detailed predictions, which made it very good but still vulnerable to chance. If they could remove this element of luck, they would have the perfect checkers player.

It might seem odd that checkers involves luck. As long as an identical series of moves are made, the game will always end with the same result. To use the mathematical term, the game is "deterministic": it is not affected by randomness like poker is. Yet when Chinook played checkers, it couldn't control the result purely through its actions, which meant that it could be beaten. In theory, it was even possible to lose to a completely incompetent opponent.

To understand why, we must look at another piece of Émile Borel's research. As well as his work on game theory, Borel was interested in very unlikely events. To illustrate how seemingly rare things will almost certainly happen if we wait long enough, he coined the infinite monkey theorem. The premise of this theorem is simple. Suppose a monkey is hammering away randomly at a typewriter (without smashing it up, as happened when a University of Plymouth team attempted this with a real monkey in 2003) and does so for an infinitely long period of time. If the monkey continues to bash away at the keys, eventually it will be almost sure to type out the complete works of Shakespeare. By sheer chance, says the theorem, the monkey will at some point hit the right letters in the order needed to reproduce all thirty-seven of the Bard's plays.

No monkey would ever live to an infinite age, let alone sit at a typewriter for that long. So, it's best to think of the monkey as a metaphor for a random letter generator, churning out an arbitrary sequence of characters. Because the letters are random, there's a chance—albeit a small one—that the first ones the monkey types are "Who's there," the opening line of *Hamlet*. The monkey might then get lucky and keep typing the correct letters until it's reproduced all of the plays. This is extremely unlikely, but it could happen. Alternatively, the monkey might type reams of utter nonsense and then finally hit good fortune with the right combination of letters. It might even type gibberish for billions of years before eventually typing the correct letters in the correct order.

Individually, each one of these events is incredibly unlikely. But because there are so many ways the monkey could end up typing out the complete works of Shakespeare—an infinite number of ways, in fact—the chances of it happening eventually are extremely high. Actually, it's almost certain to happen.

Now suppose we replaced the typewriter with a checkers board and taught our hypothetical monkey the basic rules of the game. It would therefore make a series of totally random—but valid—moves. And the infinite monkey theorem tells us that because Chinook relied on predictions, the monkey would eventually strike upon a winning combination of moves. Whereas a computer can always force a draw in tic-tac-toe, victory in checkers depended on what Chinook's opponent did. Part of the game was therefore out of its hands. In other words, winning required luck.

CHINOOK PLAYED ITS LAST competitive game in 1996. But Schaeffer and his collegues did not fully retire their champion software. Instead, they set it to work finding a checkers strategy that would never lose, no matter what the computer's opponent did. The results were finally announced in 2007, when the Alberta researchers published a paper announcing "Checkers is solved."

There are three levels of solution for a game like checkers. The most detailed, a "strong solution," describes the final outcome when perfect players pick up any game at any point, including ones where errors have already been made. This means that, whatever the starting position, we always know the optimal strategy from that point onward. Although this type of solution requires a huge amount of computation, people have found strong solutions for relatively simple games like tic-tac-toe and Connect Four.

The next type of solution is when the optimal result is known, but we only know how to reach it if we play from the start of the game. These "weak solutions" are particularly common for complicated games, where it is only feasible to look at what happens if both players make perfect moves throughout.

The most basic, an "ultraweak solution," reveals the final result when both players make a perfect sequence of moves but doesn't show what those moves are. For instance, although strong solutions have been found for Connect Four and tic-tac-toe, John Nash showed in 1949 that when any such get-so-many-in-a-row-style game is played perfectly, the player who goes second will never win. Even if we can't find the optimal strategy, we can prove this claim is true by looking at what happens if we assume that it isn't and show that our incorrect assumption leads to a logical dead-end. Mathematicians call such an approach "proof by contradiction."

To kick off our proof, let's suppose there *is* a winning sequence of moves for the second player. The first player can turn the situation to their advantage by making their opening move completely at random, waiting for the second player to reply, and then "stealing" the second player's winning strategy from that point on. In effect, the first player has turned into the second player. This "strategy-stealing" approach works because having the randomly placed extra counter on the board at the start can only improve the first player's chance of winning.

By adopting the second player's winning strategy, the first player will end up victorious. However, at the start we assumed the second player has a winning strategy. This means both players therefore win, which is clearly a contradiction. So, the only logical outcome is that the second player can never win.

Knowing that a game has an ultraweak solution is interesting but doesn't really help a player win in practice. In contrast, strong solutions, despite guaranteeing an optimal strategy, can be difficult to find when games have a lot of possible combinations of moves. Because checkers is around a million times more complicated than Connect Four, Schaeffer and colleagues focused on finding a weak solution.

When it played Marion Tinsley, Chinook made decisions in one of two ways. Early in the game, it searched through possible moves, looking ahead to see where they might lead. In the final stages, when there were fewer pieces left on the board and hence fewer possibilities to analyze, Chinook instead referred to its "endgame database" of perfect strategies. Tinsley also had a remarkable understanding of the endgame, which is partly why he was so hard to beat. This became apparent in one of his early games against Chinook in 1990. Chinook had just made its tenth move when Tinsley said, "You're going to regret that." Twenty-six moves later, Chinook resigned.

The challenge for the Alberta team was getting the two approaches to meet in the middle. In 1992, Chinook could only look ahead seventeen moves, and its endgame database only had information for situations in which there were fewer than six pieces on the board. What happened in between came down to guesswork.

Thanks to increases in computing power, by 2007 Chinook was able to search far enough into the future, and assemble a large enough endgame database, to trace out the perfect strategy from start to finish. The result, which was published in the journal *Science*, was a remarkable achievement. Yet the strategy might never have been found had it not been for the matches against Tinsley. The Alberta team later said that the Chinook project "might have died in 1990 because of a lack of human competition."

Despite it being a perfect strategy, Schaeffer wouldn't recommend using it in a game against less-skilled opponents. Chinook's early matches against humans showed that it is often beneficial to deviate from the optimal strategy if it means increasing the chances of an opponent making a mistake. This is because most players cannot see dozens of moves ahead like Chinook can. The potential for errors is even greater in games like chess and poker, where nobody knows the perfect strategy. Which raises an important question: What happens when we apply game theory to games that are too complicated to fully learn?

ALONG WITH TOBIAS GALLA, a physicist at the University of Manchester, Doyne Farmer has started to question how game theory holds up when games aren't simple. Game theory relies on the assumption that all players are rational. In other words, they are aware of the effects of the various decisions they could make, and they choose the one that benefits them most. In simple games, like tic-tac-toe or the prisoner's dilemma, it's easy to make sense of the possible options, which means players' strategies almost always end up in Nash equilibrium. But what happens when games are too complicated to fully grasp?

The complexity of chess and many forms of poker means that players, be they human or machine, haven't yet found the optimal strategy. A similar problem crops up in financial markets. Although crucial information—from share prices to bond yields—is widely available, the interactions among banks and brokers that bump and buffet the markets are too intricate to fully understand.

Poker bots try to get around the problem of complexity by "learning" a set of strategies prior to playing an actual game. But in real life, players often learn strategies *during* a game. Economists have suggested that people tend to pick strategies using "experience-weighted attraction," preferring past actions that were successful to those that were not. Galla and Farmer wondered whether this process of learning helps players find the Nash equilibrium when games are difficult. They were also curious to see what happens if the game doesn't settle down to an optimal outcome. What sort of behavior should we expect instead?

Galla and Farmer developed a game in which two computer players could each choose from fifty possible moves. Depending on which combination the two players picked, they each got a specific payoff, which had been assigned randomly before the game started. The values of these predetermined payoffs decided how competitive the game was. The payoffs varied between being zero-sum, with one player's losses equal to the other's gain, to being identical for both players. The extent of players' memory could also vary. In some games, players took account of every previous move during the learning process; in others, they put less emphasis on events further in the past.

For each degree of competitiveness and memory, the researchers looked at how players' choices changed over time as they learned to pick moves with better outcomes. When players had poor memories, the same decisions soon cropped up again and again, with players often descending into tit-for-tat behavior. But when the players both had a good memory and the game was competitive, something odd happened. Rather than settle down to equilibrium, the decisions fluctuated wildly. Like the roulette balls Farmer had attempted to track while a student, the players' choices bounced around unpredictably. The researchers found that as the number of players increased, this chaotic decision making became more common. When games are complicated, it seems that it may be impossible to anticipate players' choices.

Other patterns also emerged, including ones that had previously been spotted in real-life games. When mathematician Benoit Mandelbrot looked at financial markets in the early 1960s, he noticed that volatile periods in stock markets tended to cluster together. "Large changes tend to be followed by large changes," he noted, "and small changes tend to be followed by small changes." The appearance of "clustered volatility" has intrigued economists ever since. Galla and Farmer spotted the phenomenon in their game, too, suggesting the pattern may just be a consequence of lots of people trying to learn the complexities of the financial markets.

Of course, Galla and Farmer made several assumptions about how we learn and how games are structured. But even if real life is different, we should not ignore the results. "Even if it turns out that we are wrong," they said, "explaining why we are wrong will hopefully stimulate game theorists to think more carefully about the generic properties of real games."

ALTHOUGH GAME THEORY CAN help us identify the optimal strategy, it's not always the best approach to use when players are error-prone or have to learn. The Chinook team knew this, which is why they ensured the program picked strategies that would entice its opponents into making mistakes. Chris Ferguson was also aware of the issue. As well as employing game theory, he looked for changes in body language, adjusting his betting if players become nervous or overconfident. Players don't just need to anticipate how the perfect opponent behaves; they need to predict how *any* opponent will behave.

As we shall see in the next chapter, researchers are now delving deeper into artificial learning and intelligence. For some of them, the work has been years in the making. In 2003, an expert human player competed against one of the leading poker bots. Although the bot used game theory strategies to make decisions, it could not predict the changing behavior of its competitors. Afterward, the human player told the bot's creators, "You have a very strong program. Once you add opponent modeling to it, it will kill everyone."

7

THE MODEL OPPONENT

WHEN IT CAME TO THE GAME SHOW *JEOPARDY!* KEN JENNINGS and Brad Rutter were the best. It was 2011, and Rutter had netted the most prize money, while Jennings had gone a record seventy-four appearances without defeat. Thanks to their ability to dissect the show's famous general knowledge clues, they had won over $5 million between them.

On Valentine's Day that year, Jennings and Rutter returned for a special edition of the show. They would face a new opponent, named Watson, who had never appeared on *Jeopardy!* before. Over the course of three episodes, Jennings, Rutter, and Watson answered questions on literature, history, music, and sports. It didn't take long for the newcomer to edge into the lead. Despite struggling with the "Name the Decade" round, Watson dominated when it came to the Beatles and the history of the Olympics. Although there was a last-minute surge from Jennings, the ex-champions could not keep up. By the end of the show, Watson had racked up over $77,000, more than Jennings and Rutter combined. It was the first time Rutter had ever lost.

Watson didn't celebrate the win, but its makers did. Named after Thomas Watson, founder of IBM, the machine was the culmination of seven years of work. The idea for Watson had come during a company dinner in 2004. During the meal, an eerie silence had descended over the restaurant. Charles Lickel, IBM's research manager, realized the lack of conversation was caused by something happening on the room's television screens. Everyone was watching Ken Jennings on his phenomenal *Jeopardy!* winning streak. As Lickel looked at the screen, he realized the game could be a good test for IBM's expertise. The firm had a history of taking on human games—their Deep Blue computer had beaten chess grandmaster Garry Kasparov in 1997—but it hadn't tackled a game like *Jeopardy!* before.

To win *Jeopardy!* players need knowledge, wit, and a talent for wordplay. The show is essentially an inverted quiz. Contestants get clues about the answer and have to tell the host what the question is. So, if a clue is "5,280," the answer might be "How many feet are there in a mile?"

The finished version of Watson would use dozens of different techniques to interpret the clues and search for the correct response. It had access to the entire contents of Wikipedia and $3 million of computer processors to crunch the information.

Analyzing human language and juggling data can be useful in other less glitzy environments, too. Since Watson's victory, IBM has updated the software so that it can trawl medical databases and help with decision making in hospitals. Banks are also planning to use it to answer customer questions, while universities are hoping to employ Watson to direct student queries. By studying cookbooks, Watson is even helping chefs find new flavor combinations. In 2015, IBM collected some of the results into a "cognitive computing cookbook," which includes recipes such as a burrito with chocolate, cinnamon, and edamame beans.

Although Watson's feats on *Jeopardy!* were impressive, the show is not the ultimate test for thinking machines. There is another, arguably much bigger, challenge for artificial intelligence out there, one that predates Watson, and even Deep Blue. While Deep Blue's predecessor "Deep Thought" was gradually clambering up the chess rankings in the early 1990s, a young researcher named Darse Billings arrived at the University of Alberta. He joined the computer science department, where Jonathan Schaeffer and his team had recently developed the successful Chinook checkers program. Perhaps chess would make a good next target? Billings had other ideas. "Chess is easy," he said. "Let's try poker."

EACH SUMMER, THE WORLD'S best poker bots gather to play a tournament. In recent years, three competitors have dominated. First, there is the University of Alberta group, which currently has about a dozen researchers working on poker programs. Next, there is a team from Carnegie Mellon University in Pittsburgh, Pennsylvania, just down the road from where Michael Kent used to work while developing his sports predictions. Tuomas Sandholm, a professor in computer science, heads up the group and their work on champion bot "Tartanian." Finally, there is Eric Jackson, an independent researcher, who has created a program named "Slumbot."

The tournament consists of several different competitions, with teams tailoring their bots' personalities to each one. Some competitions are knockouts. In each round two bots go head-to-head, and the one with the smallest pile of chips at the end gets eliminated. To win these competitions, bots need strong survival instincts. They need to win only enough to get through to the next round: greed, as it were, is *not* good. In other matches, however, the winning bot is the one that gathers the most cash overall. Computer players therefore need to squeeze as much as they can out of their opponents. Bots need to go on the offensive and find ways to take advantage of the others.

Most of the bots in the competition have spent years in development, training over millions if not billions of games. Yet there are no big prize pots waiting for the winners. The creators might get bragging rights, but they won't leave with Vegas-sized rewards. So, why are these programs useful?

Whenever a computer plays poker, it is solving a problem that's familiar to all of us: how to deal with missing information. In games like chess, information is not an issue. Players can see everything. They know where the pieces are and what moves their opponent has made. Luck creeps into the game not because players can't observe events but because they are unable to process the available information. That is why there is a chance (albeit tiny) that a grandmaster could lose to a monkey picking random moves.

With a good game-playing algorithm—and a lot of computer power—it's possible to get around the information-processing problem. That's how Schaeffer and his colleagues found the perfect strategy for checkers and how a computer might one day solve chess. Such machines can beat their opponents with brute force, crunching through every possible set of moves. But poker is different. No matter how good players are, each has to cope with the fact that opponents' cards are hidden. Although the game has rules and limits, there are always unknown factors. The same problem crops up in many aspects of life. Negotiations, auctions, bargaining; they are all incomplete information games. "Poker is a perfect microcosm of many situations we encounter in the real world," Schaeffer said.

WHILE WORKING AT LOS Alamos during the Second World War, Stanislaw Ulam, Nick Metropolis, John von Neumann, and others would often play poker late into the night. The games were not particularly intense. The stakes were small, and the conversation light. Ulam said it was "a bath of refreshing foolishness from the very serious and important business that was the raison d'être of Los Alamos." During one of their games, Metropolis won ten dollars from von Neumann. He was delighted to beat a man who'd written an entire book on game theory. Metropolis used half the money to buy a copy of von Neumann's *Theory of Games and Economic Behavior* and stuck the remaining five dollars inside the cover to mark the win.

Even before von Neumann had published his book on game theory, his research into poker was well known. In 1937, von Neumann had presented his work in a lecture at Princeton University. Among the attendees, there would almost certainly have been a young British mathematician by the name of Alan Turing. At the time Turing was a graduate student visiting from the University of Cambridge. He had come to the United States to work on mathematical logic. Although he was disappointed Kurt Gödel was no longer at the university, Turing generally enjoyed his time at Princeton, even if he did find certain American habits puzzling. "Whenever you thank them for anything they say 'You're welcome,'" he told his mother in a letter. "I rather liked it at first, thinking I was welcome, but I now find it comes back like a ball against a wall, and I become positively apprehensive."

After spending a year at Princeton, Turing returned to England. Although he was based mainly in Cambridge, he also took up a part-time position with the Government Code and Cypher School at nearby Bletchley Park. When the Second World War broke out in autumn 1939, Turing found himself at the forefront of the British effort to break enemy codes. During that period, the German military encrypted radio messages using so-called Enigma machines. These typewriter-like contraptions had a series of rotors that converted keystrokes into coded text. This complexity of the encryption was a major obstacle for the code breakers at Bletchley Park. Even if Turing and his colleagues had clues about the messages—for example, certain "crib" words that were likely to appear in the text—there were still thousands of possible rotor settings to trawl through. To solve the problem, Turing designed a computer-like machine called the "bombe" to do the hard work. Once code breakers had found a crib, the bombe could identify the Enigma settings that produced the code and decipher the rest of the message.

Breaking the Enigma code was probably Turing's most famous achievement, but much like von Neumann he was also interested in games. Von Neumann's research on poker certainly grabbed Turing's attention. When Turing died in 1954, he left his friend Robin Gandy a collection of papers. Among them was a half-finished manuscript entitled "The Game of Poker," in which Turing had tried to build on von Neumann's simple analysis of the game.

Turing did not think only about the mathematical theory of games. He also wondered how games could be used to investigate artificial intelligence. According to Turing, it did not make sense to ask "can machines think?" He said the question was too vague, the range of answers too ambiguous. Rather, we should ask whether a machine is capable of behaving in a way that is indistinguishable from a (thinking) human. Can a computer trick someone into believing it is human?

To test whether an artificial being could pass for a real person, Turing proposed a game. It would need to be a fair contest, an activity that both humans and machines could succeed at. "We do not wish to penalise the machine for its inability to shine in beauty competitions," Turing said, "nor to penalise a man for losing in a race against an aeroplane."

Turing suggested the following setup. A human interviewer would talk with two unseen interviewees, one of them human and the other a machine. The interviewer would then try to guess which was which. Turing called it the "imitation game." To avoid the participants' voices or handwriting influencing things, Turing suggested that all messages be typed. While the human would be trying to help the interviewer by giving honest answers, the machine would be out to deceive its interrogator. Such a game would require a number of different skills. Players would need to process information and respond appropriately. They would have to learn about the interviewer and remember what has been said. They might be asked to perform calculations, recall facts, or tackle puzzles.

At first glance, Watson appears to fit the job description well. While playing *Jeopardy!* the machine had to decipher clues, gather knowledge, and solve problems. But there is a crucial difference. Watson did not have to play like a human to win *Jeopardy!* It played like a supercomputer, using its faster reaction times and vast databases to beat its opponents. It did not show nerves or frustration, and it didn't have to. Watson wasn't there to persuade people that it was human; it was there to win.

The same was true of Deep Blue. When it played chess against Garry Kasparov, it played in a machine-like way. It used vast amounts of computer power to search far into the future, examining potential moves and evaluating possible strategies. Kasparov pointed out that this "brute force" approach did not reveal much about the nature of intelligence. "Instead of a computer that thought and played chess like a human, with human creativity and intuition," he later said, "they got one that played like a machine." Kasparov has suggested that poker might be different. With its blend of probability and psychology and risk, the game should be less vulnerable to brute force methods. Perhaps it could even be the sort of game that chess and checkers never could, a game that needed to be learned rather than solved?

Turing saw learning as a crucial part of artificial intelligence. To win the imitation game, a machine would need to be advanced enough to pass convincingly as a human adult. Yet it did not make sense to focus only on the polished final creation. To create a working mind, it was important to understand where a mind comes from. "Instead of trying to produce a programme to simulate the adult mind," Turing said, "why not rather try to produce one which simulates the child's?" He compared the process to filling a notebook. Rather than attempting to write everything out manually, it would be easier to start with an empty notebook and let the computer work out how it should be filled.

IN 2011, A NEW type of game started appearing among the slot machines and roulette tables of Las Vegas casinos. It was an artificial version of Texas hold'em poker: the chips reduced to two dimensions, the cards dealt on a screen. In the game, players faced a single computer opponent in a two-player form of the game, commonly known as "heads-up poker."

Ever since von Neumann looked at simplified two-player games, heads-up poker has been a favorite target of researchers. This is mainly because it is much easier to analyze a game involving a pair of players than one with several opponents. The "size" of the game—measured by counting the total possible sequences of actions a player could make—is considerably smaller with only two players. This makes it much easier to develop a successful bot. In fact, when it comes to the "limit" version of heads-up poker, in which maximum bets are capped, the Vegas machines are better than most human players.

In 2013, journalist Michael Kaplan traced the origin of the machines in an article for the *New York Times*. It turned out that the poker bots owed much to a piece of software created by Norwegian computer scientist Fredrik Dahl. While studying computer science at the University of Oslo, Dahl had become interested in backgammon. To hone his skills, he created a computer program that could search for successful strategies. The program was so good that he ended up putting it on floppy disks, which he sold for $250 apiece.

Having created a skilled backgammon bot, Dahl turned his attention to the far more ambitious project of building an artificial poker player. Because poker involved incomplete information, it would be much harder for a computer to find successful tactics. To win, the machine would have to learn how to deal with uncertainty. It would have to read its opponent and weigh large numbers of options. In other words, it would need a brain.

IN A GAME LIKE poker, an action might require several decision-making steps. An artificial brain can therefore require multiple linked neurons. One neuron might evaluate the cards on display. Another might consider the amount of money on the table; a third might examine other players' bets. These neurons won't necessarily lead directly to the final decision. The results might flow into a second layer of neurons, which combine the first round of decision making in a more detailed way. The internal neurons are known as "hidden layers" because they lie between the two

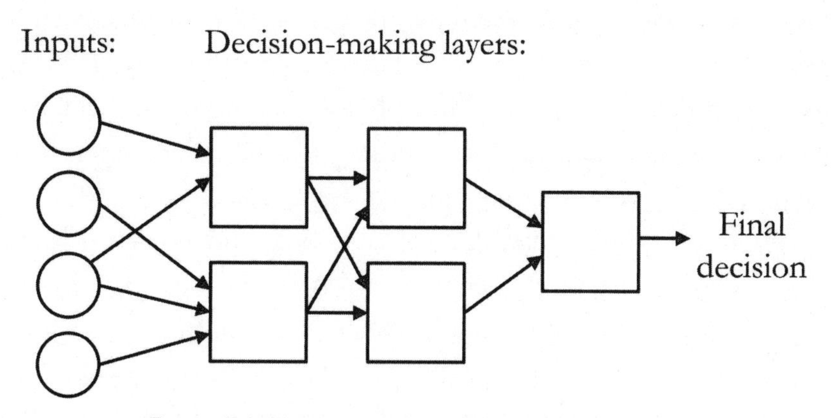

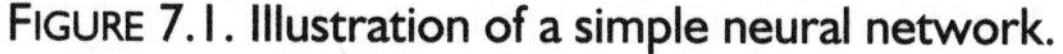
FIGURE 7.1. Illustration of a simple neural network.

visible chunks of information: what goes into the neural network and what comes out.

Neural networks are not a new idea; the basic theory for an artificial neuron was outlined in the 1940s. However, the increased availability of data and computing power means that they are now capable of some impressive feats. As well as enabling bots to learn to play games, they are helping computers to recognize patterns with remarkable accuracy.

In autumn 2013, Facebook announced an AI team that would specialize in developing intelligent algorithms. At the time, Facebook users were uploading over 350 million new photos every day. The company had previously introduced a variety of features to deal with this avalanche of information. One of them was facial recognition: the company wanted to give users the option to automatically detect—and identify—faces in their photos. In spring 2014, the Facebook AI team reported a substantial improvement in the company's facial recognition software, known as DeepFace.

The artificial brain behind DeepFace consists of nine layers of neurons. The initial layers do the groundwork, identifying where the face is in the picture and centering the image. Subsequent layers then pick out features that give a lot of clues about identity, such as the area between the eyes and eyebrows. The final neurons pull together all the separate measurements, from eye shape to mouth position, and use them to label the face. The Facebook team trained the neural network using multiple photos of four thousand different people. It was the largest facial data set ever assembled; on average, there were over a thousand pictures of each face.

With the training finished, it was time to test the program. To see how well DeepFace performed when given new faces, the team asked it to identify photos taken from "Labeled Faces in the Wild," a database containing thousands of human faces in everyday situations. The photos are a good test of facial recognition ability; the lighting isn't always the same, the camera focus varies, and the faces aren't necessarily in the same position. Even so, humans appear to be very good at spotting whether two faces are the same: in an online experiment, participants correctly matched the faces 99 percent of the time.

But DeepFace was not far behind. It had trained for so long, and had its artificial neurons rewired so many times, that it could spot whether two photos were of the same person with over 97 percent accuracy. Even when the algorithm had to analyze stills from YouTube videos—which are often smaller and blurrier—it still managed over 90 percent accuracy.

Dahl's poker program also took a long time to build experience. To train his software, Dahl set up lots of bots and got them to face off against each other in game after game. The computer programs sat through billions of hands, betting and bluffing, their artificial brains developing while they played. As the bots improved, Dahl found that they began to do some surprising things.

IN HIS LANDMARK 1952 paper "Computing Machinery and Intelligence," Turing pointed out that many people were skeptical about the possibility of artificial intelligence. One criticism, put forward by mathematician Ada Lovelace in the nineteenth century, was that machines could not create anything original. They could only do what they were told. Which meant a machine would never take us by surprise.

Turing disagreed with Lovelace, noting that "machines take me by surprise with great frequency." He generally put these surprises down to oversight. Perhaps he'd made a hurried calculation or a careless assumption while constructing a program. From rudimentary computers to high-frequency financial algorithms, this is a common problem. As we've seen, erroneous algorithms can often lead to unexpected negative outcomes.

Sometimes the error can work to a computer's advantage, however. Early in the chess match between Deep Blue and Kasparov, the machine produced a move so puzzling, so subtle, and so—well—intelligent that it threw Kasparov. Rather than grab a vulnerable pawn, Deep Blue moved its rook into a defensive position. Kasparov had no idea why it would do that. By all accounts, the move influenced the rest of the match, persuading the Russian grandmaster that he was facing an opponent far beyond anything he'd played before.

In fact, Deep Blue had no reason for choosing that particular move. Having eventually run into a situation in which it had no rules for—as predicted by Gödel's incompleteness theorem—the computer had acted randomly instead. Deep Blue's game-changing show of strategy was not an ingenious move; it was simple good luck.

Turing admitted that such surprises are still the result of human actions, with the outcomes coming from rules humans have defined (or failed to define). But Dahl's poker bots did not produce surprising actions because of human oversight. Rather, surprises were the result of the programs' learning process. During the training games, Dahl noticed that one of the bots was using a tactic known as "floating." After the three flop cards are shown, a floating player calls the opponent's bets but does not raise them. The floating player loiters, playing out the round without influencing the stakes. Once the fourth turn card is revealed, the player makes a move and raises aggressively, with the hope of scaring the opponent into folding. Dahl had not come across such a technique before, but the strategy is familiar to most good poker players. It also requires a lot of skill to pull off successfully. Not only do players need to judge the cards on display, they need to read their opponents correctly. Some are easier to scare off than others; the last thing a floating player wants is to raise aggressively and then end up in a showdown.

At first glance, such skills seem inherently human. How could a bot teach itself a strategy like this? The answer is that it is inevitable, because sometimes a play relies more on cold logic than we might think. It was just as von Neumann found with bluffing. The strategy was not a mere quirk of human psychology; it was a necessary tactic when following an optimal poker strategy.

In his *New York Times* article, Kaplan mentions that people often refer to Dahl's machine in human terms. They give it nicknames. They call it him. They even admit to talking to the metal box as if it's a real player, as if there's a person sitting behind the glass. When it comes to Texas hold'em, it appears that the bot has succeeded in making people forget that it's a computer program. If Turing's test involved a game of poker rather than a series of questions, Dahl's machine would surely pass.

Perhaps it is not particularly strange that people tend to treat poker bots as independent characters rather than viewing them as the property of the people who programmed them. After all, the best computer players are generally much better than their creators. Because the computer does all the learning, the bot doesn't need to be handed much information initially. Its human maker can therefore be relatively ignorant about game strategies, yet still end up with a strong bot. "You can do amazing things with very little knowledge," as Jonathan Schaeffer put it. In fact, despite having some of the best poker bots in the world, the Alberta poker group has limited human talent when it comes to the game. "Most of our group aren't poker players at all," researcher Michael Johanson said.

Although Dahl had created a bot that could learn to beat most players at limited-stakes poker, there was a catch. Las Vegas gaming rules stipulate that gaming machines have to behave the same against all players. They can't tailor their playing style for opponents who are skilled or inexperienced. The rules meant that Dahl's bot had to sacrifice some of its cunning before it was allowed on the casino floor. From a bot's point of view, having to follow a fixed strategy can make things more difficult. Having a rigid adult brain—rather than the flexible one of a child—prevents the machine from learning how to exploit weaknesses. This removes a big advantage, because it turns out that humans have plenty of flaws that can be exploited.

IN 2010, AN ONLINE version of rock-paper-scissors appeared on the *New York Times* website. It's still there if you want to try it. You'll be able to play against a very strong computer program. Even after a few games, most people find that the computer is pretty hard to beat; play lots of games, and the computer will generally end up in the lead.

Game theory suggests that if you follow the optimal strategy for rock-paper-scissors, and choose randomly between the three available options, you should expect to come out even. But when it comes to rock-paper-scissors, it seems that humans aren't very good at doing what's optimal. In 2014, Zhijian Wang and colleagues at Zhejiang University in China reported that people tend to follow certain behavior patterns during games of rock-paper-scissors. The researchers recruited 360 students, divided them into groups, and asked each group to play three hundred rounds of rock-paper-scissors against each other. During the games, the researchers found that many students adopted what they called a "win-stay lose-shift" strategy. Players who'd just won a round would often stick with the same action in the next round, while the losing players had a habit of switching to the option that beat them. They would swap rock for paper, for instance, or scissors for rock. Over many rounds, the players generally chose the three different options a similar number of times, but it was clear they weren't playing randomly.

The irony is that even truly random sequences can contain seemingly nonrandom patterns. Remember those lazy journalists in Monte Carlo who made up the roulette numbers? There were a lot of obstacles they'd have had to overcome to create results that appeared random. First, they would have had to make sure black and red came up similarly often in the results. The journalists actually managed to get this bit right, which meant the data passed the initial round of Karl Pearson's "Is it random?" test. However, the reporters came unstuck when it came to runs of colors, because they switched between red and black more often than a truly random sequence would.

Even if you know how randomness should look, and try to alternate between colors—or rock, paper, scissors—correctly, your ability to generate random patterns will be limited by your memory. If you had to read a list of numbers and immediately recite them, how many could you manage? Half a dozen? Ten? Twenty?

In the 1950s, cognitive psychologist George Miller noted that most young adults could learn and recite around seven numbers at a time. Try memorizing one local phone number and you'll probably be fine; attempt to remember two, and it gets tricky. This can be problematic if you're trying to generate random moves in a game; how can you ensure you use all options equally often if you can only remember the last few moves? In 1972, Dutch psychologist Willem Wagenaar observed that people's brains tend to concentrate on a moving "window" of about six to seven previous responses. Over this interval, people could alternate between options reasonably "randomly." However, they were not so good at switching between options over longer time intervals. The size of the window, around six to seven events long, could well be a consequence of Miller's earlier observation.

In the years since Miller published his work, researchers delved further into human memory capacity. It turns out that the value Miller jokingly referred to as the "magical number seven" is not so magical after all. Miller himself noted that when people had to remember only binary numbers—such as zeros and ones—they could recite a sequence of about eight digits. In fact, the size of the data "chunks" humans can remember depends on the complexity of the information. People might be able to recall seven numbers, but there is evidence they can recite only six letters or so, and five one-syllable words.

In some cases, people have learned to increase the amount of information they can recall. In memory championships, the best competitors can memorize over a thousand playing cards in an hour. They do this by changing the format of the data chunks they remember; rather than thinking in terms of raw numbers, they try to memorize images as part of a journey. Cards become celebrities or objects; the sequence becomes a series of events in which their card characters feature. This helps the competitors' brains shelve and retrieve the information more efficiently. As discussed in the previous chapter, memorizing cards also helps in blackjack, with card counters "bucketing" information to reduce the amount they have to store. Such storage problems have interested researchers looking at artificial minds as well as those working on human ones. Nick Metropolis said Stanislaw Ulam "often mused about the nature of memory and how it was implemented in the brain."

When it comes to rock-paper-scissors, machines are much better than humans at coming up with the unpredictable moves required for an optimal game theory strategy. Such a strategy is inherently defensive, of course, because it aims to limit potential losses against a perfect opponent. But the rock-paper-scissors bot on the *New York Times* website was not playing a perfect opponent. It was playing error-prone humans, who have memory issues and can't generate random numbers. The bot therefore deviated from a random strategy and started to hunt down weaknesses.

The computer had two main advantages over its human opponents. First, it could accurately remember what the human had done in previous rounds. It could recall which sequences of moves the person had played, for example, and which patterns the person was fond of. And that's when the second advantage kicked in.

The computer wasn't just using information on its current opponent. It was drawing on knowledge gained during two hundred thousand rounds of rock-paper-scissors against humans. The database came from Shawn Bayern, a law professor and an ex-computer scientist whose website runs a massive online rock-paper-scissors tournament. The competition is still going, with over half a million rounds played to date (the computer has won the majority of them). The data meant the bot could compare its current opponent with others it had played. Given a particular sequence of moves, it could work out what humans tended to do next. Rather than being interested only in randomness, the machine was instead building a picture of its opponent.

Such approaches can be particularly important in games like poker, which can have more than two players. Recall that, in game theory, optimal strategies are said to be in Nash equilibrium: no single player will gain anything by picking a different strategy. Neil Burch, one of the researchers in the University of Alberta poker group, points out that it makes sense to look for such strategies if you have a single opponent. If the game is zero-sum—with everything you lose going to your opponent, and vice versa—then a Nash equilibrium strategy will limit your losses. What's more, if your opponent deviates from an equilibrium strategy, your opponent will lose out. "In two player games that are zero-sum, there's a really good reason to say that a Nash equilibrium is the correct thing to play," Burch said. However, it isn't necessarily the best option when more players join the game. "In a three player game, that can fall apart."

Nash's theorem says players will lose out if they change their strategy unilaterally. But it doesn't say what will happen if two players swap tactics together. For instance, two of the players could decide to gang up on the third. When von Neumann and Morgenstern wrote their book on game theory, they noted that such coalitions work only when there are at least three players. "In a two-person game there are not enough players to go around," they said. "A coalition absorbs at least two players, and then nobody is left to oppose." Turing also acknowledged the potential role of coalitions in poker. "It is only etiquette, sense of fair play etc. which prevents this happening in actual games," he said.

There are two main ways to form coalitions in poker. The most blatant way to collude would be for two or more players to reveal their cards to each other. When one of them received a strong hand, they all could push the bets up gradually to squeeze more money out of their opponents. Naturally, this approach is much easier to employ while playing online. Parisa Mazrooei and colleagues at the University of Alberta suggest such collusion should really be referred to as "cheating" because players are using strategies outside the rules of the game, which stipulate that cards remain hidden.

The alternative would be for colluders to keep their cards to themselves but give signals to other players when they had a strong hand. Technically, they would be operating within the rules (if not inside the boundaries of fair play). Colluding players often follow certain betting patterns to improve their chances. If one player bets big, for example, the others follow to drive opponents out of the game. Human players would have to remember such signals, but things are easier for bots, which can have access to the exact set of programmed rules used by fellow colluders.

There are reports of unscrupulous players using both types of approach in online poker rooms. However, it can be difficult to detect collusion. If a player is matching another's bets, gradually inflating the pot, that player might be manipulating the game to help their teammate. Or the person could be just a naive beginner, trying to bluff to victory. "In any form of poker, there exist a large variety of strategy combinations that are mutually beneficial to those that apply them," Frederik Dahl has pointed out. "If they apply such strategies on purpose, we would say that they cheat by co-operating, but if it happens just by accident, we would not."

That's the problem with using game theory in poker: coalitions don't always have to be deliberate. They might just result from the strategies players choose. In many situations, there is more than one Nash equilibrium. Take driving a car. There are two equilibrium strategies: if everyone drives on the left, you will lose out if you make a unilateral decision to drive on the other side; if driving on the right is in vogue, the left is no longer the best choice.

Depending on the location of your driver's seat, one of these equilibriums will be preferable to the other. If your car is left-hand drive, for instance, you'll probably prefer it if everyone drives on the right. Obviously, having a driver's seat inconveniently positioned on the "wrong" side of the car won't be enough to make you change the side you drive on. But the situation is still a bit like having everyone else in a coalition against you (if you're feeling particularly grumpy about things). Drive on the other side of the road and you'll clearly lose out, so you just have to put up with the situation.

The same problem crops up in poker. As well as causing inconvenience, it can cost players money. Three poker players could choose Nash equilibrium strategies, and when these strategies are put together, it may turn out that two players have selected tactics that just so happen to pick on the third player. This is why three-player poker is so difficult to tackle from a game theory point of view. Not only is the game far more complicated, with more potential moves to analyze, it's not clear that hunting for the Nash equilibrium is always the best approach. "Even if you could compute one," Michael Johanson said, "it wouldn't necessarily be useful."

There are other drawbacks, too. Game theory can show you how to minimize your losses against a perfect opponent. But if your opponent has flaws—or if there are more than two players in the game—you might want to deviate from the "optimal" Nash equilibrium strategy and instead take advantage of weaknesses. One way to do this would be to start off with an equilibrium strategy, and then gradually tweak your tactics as you learn more about your opponent. Such approaches can be risky, however. Tuomas Sandholm at Carnegie Mellon University points out that players must strike a balance between exploitation and exploitability. Ideally, you want to *exploit*, taking as much as possible from weak opponents, but not be *exploitable*, and come unstuck against strong players. Defensive strategies—such as the Nash equilibrium, and the tactics employed by Dahl's poker bot—are not very exploitable. Strong players will struggle to beat them. However, this comes at the price of not exploiting weak opponents much; bad players will get off lightly. It would therefore make sense to vary strategy depending on the opponent. As the old saying goes, "Don't play the cards, play the person."

Unfortunately, learning to exploit opponents can in turn leave players vulnerable to exploitability. Sandholm calls it the "get taught and exploited problem." For example, suppose your opponent appears to play aggressively at first. When you notice this, you might adjust tactics to try to take advantage of this aggression. However, at this point the opponent could suddenly become more conservative and exploit the fact that you believe—incorrectly—that you're facing an aggressive player.

Researchers can judge the effects of such problems by measuring the exploitability of their bots. This is the maximum amount they could expect to lose if the bot had made completely the wrong assumptions about its opponent. Along with PhD student Sam Ganzfried, Sandholm has been developing "hybrid" bots, which combine defensive Nash equilibrium tactics with opponent modeling. "We would like to only attempt to exploit weak opponents," they said, "while playing the equilibrium against strong opponents."

IT'S CLEAR THAT POKER programs are getting better and better. Every year, the bots in the Annual Computer Poker Competition become smarter, and Vegas is filling up with poker machines that can beat most casino visitors. But have computers truly overtaken humans? Are the best bots really better than all people?

According to Sandholm, it's hard to say whether the crossover has happened for several reasons. To start with, you'd have to identify who the best human is. Unfortunately, it's hard to rank players definitively: poker doesn't have a clear Garry Kasparov or Marion Tinsley. "We don't really know who the best human is," Sandholm said. Games against humans are also difficult to arrange. Although there is a computer competition every year, Sandholm points out that mixed matches are far less common. "It's hard to get pros to do these man-machine matches."

There has been the occasional match-up. In 2007, professional players Phil Laak and Ali Eslami took on Polaris, a bot created by the University of Alberta group, in a series of two-player poker games. Polaris was designed to be hard to beat. Rather than attempting to exploit its opponents, it employed a strategy that was close to Nash equilibrium.

At the time, some of the poker community thought Laak and Eslami were strange choices for the match. Laak had a reputation for being hyperactive at the poker table, jumping around, rolling on the floor, doing push-ups. In contrast, Eslami was fairly unknown, having appeared in relatively few televised tournaments. But Laak and Eslami had skills that the researchers needed. Not only were they good players, they were able to say what they were thinking during a game, and they were comfortable with the unusual setup involved in a man-versus-machine match.

The venue for the match was an artificial intelligence conference in Vancouver, Canada, and the game was limit Texas hold'em: the same game Dahl's bot would later play in Vegas. Although Laak and

Eslami would play Polaris in separate games, their scores would be combined at the end of each session. It was to be humans versus the machine, with Laak and Eslami playing as a team against Polaris. To minimize the effects of luck, the card deals were mirrored: whichever cards were dealt to Polaris in one game, the human player got in the other (and vice versa). The organizers also imposed a clear margin of victory. To win, a team needed to finish with at least $250 more chips than their opponent.

On the first day, there were two sessions of play, each consisting of five hundred hands. The first session ended as a draw (Polaris had finished seventy dollars up, which was not enough to be considered a victory). In the second session, Laak was lucky enough to be dealt good cards against Polaris, which meant the computer received the same strong cards in the game against Eslami. Polaris capitalized on the advantage more than Laak did, and the bot ended the day with a clear win over the human team.

That night, Laak and Eslami met up to discuss the thousand poker hands they'd just played. The Alberta team gave them the logbook of the day's play, including all the hands that had been dealt. This helped the pair dissect the games they'd played. When they returned to the tables the next day, the humans had a much better idea of how to tackle Polaris, and they won the final two sessions. Even so, the humans were modest about their victory. "This was not a win for us," Eslami said at the time. "We survived. I played the best heads-up poker I've ever played and we just narrowly won."

The following year, there was a second man-machine competition, with a new set of human opponents. This time, seven human players would take on the University of Alberta's bot in Las Vegas. The humans were arguably some of the best around; several had career winnings totaling more than $1 million. But they would not be facing the same Polaris that lost the previous year. This was Polaris 2.0. It was more advanced and better trained. Since the games against Laak and Eslami, Polaris had played over eight billion games against itself. It was now better at exploring the vast combination of possible moves, which meant there were fewer weak links in its strategy for opponents to target.

Polaris 2.0 also put more emphasis on learning. During a match, the bot would develop a model of its opponent. It would identify which type of strategy a player was using and tailor its game to target weaknesses. Players couldn't beat Polaris by discussing tactics between games as Laak and Eslami did, because Polaris would be playing differently against each one of them. Nor could humans regain the advantage by altering their own playing style. If Polaris noticed its opponent had changed strategy, the bot would adapt to the new tactics. Michael Bowling, who headed up the Alberta team, said that many of the human players struggled against Polaris's new box of tricks; they had never seen an opponent switch strategy like that.

As before, the players paired up to take on Polaris in the limit version of Texas hold'em. There were four matches in total, spread over four days. The first two went badly for Polaris, with one a draw and the other a victory for the humans. But this time the humans did not finish strongly; Polaris won the final two matches, and with them the competition.

Whereas Polaris 2.0 drifted away from an optimal strategy to exploit its opponents, the next challenge for the Alberta team was how to make a bot that was truly unbeatable. Their existing bots could only compute an approximate Nash equilibrium, which meant there might have been a strategy out there that could beat them. Bowling and his colleagues therefore set out to find a set of flawless tactics, which in the long run would not lose money against any opponent.

Using the regret minimization approach we encountered in the previous chapter, the Alberta researchers honed the bots, and then got them to play each other again and again, at a rate of about two thousand games per second. Eventually, the bots learned how to avoid being exploited, even by a perfect player. In 2015, the team unveiled their unbeatable poker program—named Cepheus—in the journal *Science*. With a nod to the group's checkers research, the paper was titled "Heads-Up Limit Hold'em Poker Is Solved."

Some of the findings lined up with conventional wisdom. The team showed that in heads-up poker, the dealer—who goes second—holds the advantage as well as the cards. They also found that Cepheus rarely "limps" and chooses to raise or fold on its first go rather than simply calling its opponent's bet. According to Johanson, as the bot narrowed in on the optimal strategy it also started to come up with some unexpected tactics. "Every now and then we find differences between what the program chooses and what human wisdom says." For example, the final version of Cepheus chooses to play hands—such as a four and six with different suits—that many humans would fold. In 2013, the team also noticed their bot would occasionally place the minimum allowable bet rather than putting down a larger stake. Given the extent of the bot's training, this was apparently the optimal thing to do. But Burch points out that human players would see things differently. Although the computer had decided it is a smart tactic, most humans would view it as annoying. "It's almost a nuisance bet," Burch said. The polished version of Cepheus is also reluctant to bet big initially. Even when it has the best hand (a pair of aces), it will bet the maximum possible amount only 0.01 percent of the time.

Cepheus has shown that, even in complex situations, it can be possible to find an optimal strategy. The researchers point to a range of scenarios in which such algorithms could be useful, from designing coast guard patrols to medical treatments. But this was not the only reason for the research. The team ended their *Science* paper with a quote from Alan Turing: "It would be disingenuous of us to disguise the fact that the principal motive which prompted the work was the sheer fun of the thing."

Despite the breakthrough, not everyone was convinced that it represented the ultimate victory of the artificial over the biological. Michael Johanson says that many human players view limit poker as the easy option, because there is a cap on how much players can raise the betting. It means the boundaries are well defined, the possibilities constrained.

No-limit poker is seen as the bigger challenge. Players can raise by any amount and can go all in whenever they want. This creates more options, and more subtleties. The game therefore has a reputation for being more of an art than a science. And that's why Johanson would love to see a computer win. "It would attack the mystique that poker is all about psychology," he said, "and that computers couldn't possibly do it."

Sandholm says it won't be long before two-player no-limit poker has fallen to the machines. "We're working on that very actively," he said. "We may already have bots that are better than the best pros." Indeed, Carnegie Mellon's bot Tartanian put in a very strong showing in the 2014 computer poker competition. There were two types of contest for no-limit poker, and Tartanian came out on top in both. As well as winning the knockout competition, it prevailed in the total bankroll contest. Tartanian could survive when it needed to but also rack up plenty of chips against weaker opponents.

As bots get better—and beat more humans—players could end up learning poker from machines. Chess grandmasters already use computers to hone their skills in training. If they want to know how to play a particularly tricky position, the machines can show them the best way to proceed. Chess computers can come up with strategies that stretch further into the future than we humans can manage.

With computer programs cleaning up at chess, checkers, and now poker, it might be tempting to argue that humans can no longer compete at such games. Computers can analyze more data, remember more strategies, and examine more possibilities. They are able to spend longer learning and longer playing. Bots can teach themselves supposedly "human" tactics such as bluffing and even "superhuman" strategies humans haven't spotted yet. So, is there anything that computers are not so good at?

ALAN TURING ONCE NOTED that if a man tried to pretend to be a machine, "he would clearly make a very poor showing." Ask the human to perform a calculation, and he'd be much slower, not to mention more error prone, than the computer. Even so, there are still some situations that bots struggle with. When playing *Jeopardy!* Watson found the short clues the most difficult. If the host read out a single category and a name—such as "first ladies" and Ronald Reagan—Watson would take too long to search through its database to find the correct response (which is "Who is Nancy Reagan?"). Whereas Watson would beat a human contestant in a race to solve a long, complicated clue, the human would prevail if there were only a few words to go by. In quiz shows, it seems that brevity is the enemy of machines.

The same is true of poker. Bots need time to study their opponents, learning their betting style so it can be exploited. In contrast, human professionals are able to evaluate other players much more quickly. "Humans are good at making assumptions about an opponent with very little data," Schaeffer said.

In 2012, researchers at the University of London suggested that some people might be especially good at sizing up others. They designed a game, called the Deceptive Interaction Task, to test players' ability to lie and detect lies. In the task, participants were placed in groups, with one person given a cue card containing an opinion—such as "I'm in favor of reality TV"—and instructions to either lie or tell the truth. After stating the opinion, the person had to give reasons for holding that view. The others in the group had to decide whether they thought the person was lying or not.

The researchers found that people who were lying generally took longer to start to speak after receiving the cue card. Liars took an average 6.5 seconds, compared to 4.6 seconds for honest speakers. It also turned out that good liars were also effective lie detectors, much like the proverb "it takes a thief to catch a thief." Although liars appeared to be better at spotting deceit in the game, it was not clear why this was the case. The researchers suggested it might be because they were better—whether consciously or unconsciously—at picking up on others' slow responses as well as speeding up their own speech.

Unfortunately, people aren't so good at identifying the specific signs of lying. In a 2006 survey spanning fifty-eight countries, participants were asked "How can you tell when people are lying?" One answer dominated the responses, coming up in every country and topping the list in most: liars avoid eye contact. Although it's a popular lie-detection method, it doesn't appear to be a particularly good one. There's no evidence that liars avert their gaze more than truthful people. Other supposed giveaways have dubious foundations, too. It is not clear that liars are noticeably more animated or shift posture when speaking.

Behavior might not always reveal liars, but it can influence games in other ways. Psychologists at Harvard University and Caltech have shown that having certain facial expressions can lure opponents into making bad bets. In a 2010 study, they got participants to play a simplified poker game against a computer-generated player, whose face was displayed on a screen. The researchers told participants the computer would be using different styles of play but said nothing about the face on the screen. In reality, the instructions were a ruse. The computer picked moves randomly; all that changed was its face. The simulated player displayed three possible expressions, which followed stereotypes about honesty. One was seemingly trustworthy, one neutral, and one untrustworthy. The researchers found that players facing computer players with dishonest or neutral faces made relatively good choices. However, when they played "trustworthy" computer opponents, participants made significantly worse decisions, often folding when they had the stronger hand.

The researchers pointed out that the study involved a cartoon version of poker, played by beginners. Professional poker games are likely to be very different. However, the study suggests that facial expressions might not influence poker in the way we assume. "Contrary to the popular belief that the optimal poker face is neutral in appearance," the authors noted, "the face that invokes the most betting mistakes by our subjects has attributes that are correlated with trustworthiness."

Emotion can also influence overall playing style. The University of Alberta poker group has found that humans are particularly susceptible to strong-arm tactics. "In general, a lot of the knowledge that human poker pros have about how to beat other humans revolves around aggression," Michael Johanson said. "An aggressive strategy that puts a lot of pressure on opponents, making them make tough decisions, tends to be very effective." When playing humans, the bots try to mimic this behavior and push opponents into making mistakes. It seems that bots have a lot to gain by copying the behavior of humans. Sometimes, it even pays to copy their flaws.

WHEN MATT MAZUR DECIDED to build a poker bot in 2006, he knew it would have to avoid detection. Poker websites would ban anyone they suspected of running computer players. It wasn't enough to have a bot that could beat humans; Mazur would need a bot that could look human while doing it.

A computer scientist based in Colorado, Mazur worked on a variety of software projects in his spare time. In 2006, the new project was poker. Mazur's first attempt at a bot, created that autumn, was a program that played a "short stacking" strategy. This involved buying into games with very little money, and then playing very aggressively, hoping to scare off players and steal the pot. It's often seen as an irritating tactic, and Mazur discovered it wasn't a particularly successful one either. Six months in, the bot had played almost fifty thousand hands and lost over $1,000. Abandoning his flawed first draft, Mazur designed a new bot, which would play two-player poker properly. The finished bot played a tight game, choosing its moves carefully, and was aggressive in its betting. Mazur said the bot was reasonably competitive against humans in small-stakes games.

The next challenge was to avoid getting caught. Unfortunately, there wasn't much information out there to help Mazur. "Online poker sites are understandably quiet when it comes to what they look at to detect bots," he said, "so bot developers are forced to make educated guesses." While designing his poker program, Mazur therefore tried to put himself in the position of a bot hunter. "If I was trying to detect a bot, I would look at a lot of different factors, weigh them, and then manually investigate the evidence in order to make a call as far as whether a player was a bot or not."

One obvious red flag would be strange betting patterns. If a bot placed too many bets, or too quickly, it might look suspicious. Unfortunately, Mazur found his bots could sometimes behave strangely by accident. The bots worked in pairs to compete on poker websites. One of them would register for new games, and the other would play them. On one occasion, Mazur was away from his computer when the game-playing program crashed. The other bot had no idea what had happened, so it kept on registering for new games. Without the game-playing bot ready to take a seat at the table, Mazur's account skipped over twenty games in a row. Mazur later realized his bots had other quirks, too. For instance, they would often play with the same stakes for hundreds of games. Mazur points out that humans rarely behave like that: they would generally get confident (or bored) over time and move up to higher-stakes games for a while.

As well as playing sensibly, Mazur's bots also had to navigate their way around the poker websites. Mazur found that some websites had features—be they accidental or deliberate—that made automated navigation harder. Sometimes they would subtly alter what appeared on his screen, perhaps by changing the size or shape of windows or moving buttons. Such changes wouldn't cause problems for a human, but they could throw bots that had been taught to navigate a specific set of dimensions. Mazur had to get his bots to track the locations of the windows and buttons and adjust where they clicked to account for any changes.

The whole process was like a version of Turing's imitation game. To avoid detection, Mazur's bots had to convince the website they were playing like humans. Sometimes, bots even found themselves facing Turing's original test. Most poker websites include a chat feature, which lets players talk to each other. Generally, this isn't a problem; players often remain silent in poker games. But there were some conversations that Mazur decided he couldn't avoid. If someone accused his bot of being a computer program and the bot didn't reply, there was a risk that he would be reported to the website owners. Mazur therefore put together a list of terms that suspicious opponents might use. If someone mentioned words such as "bot" or "cheater" during a game, he'd get an alert and intervene. It meant he'd have to be near his computer when his bot was playing, but the alternative was potentially much worse; an unsupervised program could easily run into trouble and not know to get out of it.

It took a while for Mazur's bots to become winners: the programs didn't make money for the first eighteen months they were active. Eventually, in spring 2008, the bots started to produce modest profits. The successful run came to an abrupt end a few months later, however. On October 2, 2008, Mazur got an e-mail from the poker website informing him that his account had been suspended. So, what gave it away? "In retrospect," he said, "I think the thing that got my bot caught was that it was simply playing too many games." Mazur's bot concentrated on heads-up "Sit 'n Go" games, which commence as soon as two players join the game. "A normal player might play ten to fifteen No Limit Heads Up Sit 'n Gos in a day," Mazur said. "At its peak, my bot was playing fifty to sixty per day. That probably threw up some flags." Of course, this is only his best guess. "It's possible that it was something else entirely. I'll probably never know for sure."

Mazur wasn't actually that bothered about the loss of profit from his bot. "When my account was eventually suspended, I had not netted that much money," he said. "I would have been much better off financially if I'd actually used that time to play poker instead. But then again, I didn't build the bot to make money; I built it for the challenge."

After his account was suspended, Mazur e-mailed the poker website that had banned him and offered to explain exactly what he'd done. He knew several ways to make life even harder for bots, which he hoped might improve security for human poker players. Mazur told the company all the things they should look out for, from high volumes of games to unusual mouse movements. He even suggested countermeasures that could hinder bot development, such as varying the size and location of buttons on the screen.

Mazur also posted a detailed history of his bot's creation on his website, including screenshots and schematics. He wanted to show people that poker bots are hard to build, and there are much more useful things they could be doing with computers. "I realized that if I was going to spend that much time on a software project, I should devote that energy to more worthwhile endeavors." Looking back, however, he doesn't regret the experience. "Had I not built the poker bot, who knows where I'd be."

8

BEYOND CARD COUNTING

If you ever visit a Las Vegas casino, look up. Hundreds of cameras cling to the ceiling like jet-black barnacles, watching the tables below. The artificial eyes are there to protect the casino's income from the quick-witted and light-fingered. Until the 1960s, casinos' definition of such cheating was fairly clear-cut. They only had to worry about things like dealers paying out on losing hands or players slipping high-value chips into their stake after the roulette ball had landed. The games themselves were fine; they were unbeatable.

Except it turned out that wasn't true. Edward Thorp found a loophole in blackjack big enough to fit a best-selling book through. Then a group of physics students tamed roulette, traditionally the epitome of chance. Beyond the casino floor, people have even scooped lottery jackpots using a mix of math and manpower.

The debate over whether winning depends on luck or skill is now spreading to other games. It may even determine the fate of the once lucrative American poker industry. In 2011, US authorities shut down a number of major poker websites, bringing an end to the "poker boom" that had gripped the country for the previous few years. The legislative muscle for the shake-up came from the Unlawful Internet Gambling Enforcement Act. Passed in 2006, it banned bank transfers related to games where the "opportunity to win is predominantly subject to chance." Although the act has helped curb the spread of poker, it doesn't cover stock trading or horseracing. So, how do we decide what makes something a game of chance?

During the summer of 2012, the answer would turn out to be worth a lot to one man. As well as taking on the big poker companies, federal authorities had also gone after people operating smaller games. That included Lawrence DiCristina, who ran a poker room on Staten Island in New York. The case went to trial in 2012, and DiCristina was convicted of operating an illegal gambling business.

DiCristina launched a motion to dismiss the conviction, and the following month he was back in court arguing his case. During the hearing, DiCristina's lawyer called economist Randal Heeb as an expert witness. Heeb's aim was to convince the judge that poker was predominantly a game of skill, and therefore didn't fall under the definition of illegal gambling. While giving evidence, Heeb presented data from millions of online poker games. He showed that, bar a few bad days, the top-ranked players won pretty consistently. In contrast, the worst players lost throughout the year. The fact that some people could make a living from poker was surely evidence that the game involved skill.

The prosecution also had an expert witness, an economist named David DeRosa. He did not share Heeb's views about poker. DeRosa had used a computer to simulate what might happen if a thousand people each tossed a coin ten thousand times. Assuming a certain outcome—such as tails—was equivalent to a win, and the number of times a particular person won the toss was totally random. And yet the results that came out were remarkably similar to those Heeb presented: a handful of people appeared to win consistently, and another group of people seemed to lose a large number of times. This wasn't evidence that a coin toss involves skill, just that—much like the infinite number of monkeys typing—unlikely events can happen if we look at a large enough group.

Another concern for DeRosa was the number of players who lost money. Based on Heeb's data, it seemed that about 95 percent of people playing online poker ended up out of pocket. "How could it be skillful playing if you're losing money?" DeRosa said. "I don't consider it skill if you lose less money than the unfortunate fellow who lost more money."

Heeb admitted that, in a particular game, only 10 to 20 percent of players were skillful enough to win consistently. He said the reason so many more people lost than won was partly down to the house fee, with poker operators taking a cut from the pot of money in each round (in DiCristina's games, the fee was 5 percent). But he did not think the apparent existence of a skilled poker elite was the result of chance. Although a small group may appear to win consistently if lots of people flip coins, good poker players generally continue to win after they've been ranked highly. The same cannot be said for the people who are fortunate with coin tosses.

According to Heeb, part of the reason good players can win is that in poker players have control over events. If bettors place a bet on a sports match or a roulette wheel, their wagers do not affect the result. But poker players can change the outcome of the game with their betting. "In poker, the wager is not in the same sense a wager on the outcome," Heeb said. "It is the strategic choice that you are making. You are trying to influence the outcome of the game."

But DeRosa argued that it doesn't make sense to look at a player's performance over several hands. The cards that are dealt are different each time, so each hand is independent of the last. If a single hand involves a lot of luck, there is no reason to think that player will have a successful round after a costly one. DeRosa compared the situation to the Monte Carlo fallacy. "If red has come up 20 times in a row in roulette," he said, "it does not mean that 'black is due.'"

Heeb conceded that a single hand involves a lot of chance, but it did not mean the game was chiefly one of luck. He used the example of a baseball pitcher. Although pitching involves skill, a single pitch is also susceptible to chance: a weak pitcher could produce a good ball, and a strong pitcher could throw a bad one. To identify the best—and worst—pitchers, we need to look at lots of throws.

The key issue, Heeb argued, is how long we must wait for the effects of skill to outweigh chance. If it takes a large number of hands (i.e., longer than most people will play), then poker should be viewed as a game of chance. Heeb's analysis of the online poker games suggested this wasn't the case. It seemed that skill overtook luck after a relatively small number of hands. After a few sessions of play, a skillful player could therefore expect to hold an advantage.

It fell to the judge, a New Yorker named Jack Weinstein, to weigh the arguments. Weinstein noted that the law used to convict DiCristina—the Illegal Gambling Business Act—listed games such as roulette and slot machines, but it did not explicitly mention poker. Weinstein said it wasn't the first time a law had failed to specify a crucial detail. In October 1926, airport operator William McBoyle helped arrange the theft of an airplane in Ottawa, Illinois. Although he was convicted under the National Motor Vehicle Theft Act, McBoyle appealed the result. His lawyers argued that the act did not explicitly cover airplanes, because it defined a vehicle as "an automobile, automobile truck, automobile wagon, motor cycle, or any other self-propelled vehicle not designed for running on rails." According to McBoyle's lawyers, this meant an airplane was not a vehicle, and so McBoyle could not be guilty of the federal crime of transporting a stolen vehicle. The US Supreme Court agreed. They noted that the wording of the law evoked the mental image of vehicles moving on land, so it shouldn't be extended to aircraft simply because it seemed that a similar rule ought to apply. The conviction was reversed.

Although poker wasn't mentioned in the Gambling Act, Judge Weinstein said this didn't automatically mean the game wasn't gambling. But the omission did mean that the role of chance in poker was up for debate. And Weinstein had found Heeb's evidence convincing. Until that summer, no court had ever ruled on whether poker was gambling under federal law. Weinstein delivered his conclusion on August 21, 2012, and ruled that poker was predominantly governed by skill rather than chance. In other words, it did not count as gambling under federal law. DiCristina's conviction was overturned.

The victory was to be short-lived, however. Although Weinstein ruled that DiCristina had not broken federal law, New York State has a stricter definition of gambling. Its laws cover any game that "depends in a material degree upon an element of chance." As a result, DiCristina's acquittal was overturned in August 2013. Weinstein's ruling on the relative role of luck and skill was not questioned. Rather, the state law meant that poker still fell under the definition of a gambling business.

The DiCristina case is part of a growing debate about how much luck comes into games like poker. Definitions like "material degree of chance" will undoubtedly raise more questions in the future. Given the close links between gambling and certain parts of finance, surely this definition would cover some financial investments, too? Where do we draw the line between flair and fluke?

IT IS TEMPTING TO sort games into separate boxes marked luck and skill. Roulette, often used as an example of pure luck, might go into one; chess, a game that many believe relies only on skill, might go in the other. But it isn't this simple. To start with, processes that we think are as good as random are usually far from it.

Despite its popular image as the pinnacle of randomness, roulette was first beaten with statistics, and then with physics. Other games have fallen to science too. Poker players have explained game theory and syndicates have turned sports betting into investments. According to Stanislaw Ulam, who worked on the hydrogen bomb at Los Alamos, the presence of skill is not always obvious in such games. "There may be such a thing as habitual luck," he said. "People who are said to be lucky at cards probably have certain hidden talents for those games in which skill plays a role." Ulam believed the same could be said of scientific research. Some scientists ran into seemingly good fortune so often that it was impossible not to suspect that there was an element of talent involved. Chemist Louis Pasteur put forward a similar philosophy in the nineteenth century. "Chance favours the prepared mind" was how he put it.

Luck is rarely embedded so deeply in a situation that it can't be altered. It might not be possible to completely remove luck, but history has shown that it can often be replaced by skill to some extent. Moreover, games that we assume rely solely on skill do not. Take chess. There is no inherent randomness in a game of chess: if two players make identical moves every time, the result will always be the same. But luck still plays a role. Because the optimal strategy is not known, there is a chance that a series of random moves could defeat even the best player.

Unfortunately, when it comes to making decisions, we sometimes take a rather one-sided view of chance. If our choices do well, we put it down to skill; if they fail, it's the result of bad luck. Our notion of skill can also be skewed by external sources. Newspapers print stories about entrepreneurs who have hit a trend and made millions or celebrities who have suddenly become household names. We hear tales of new writers who have produced instant best sellers and bands that have become famous overnight. We see success and wonder why those people were so special. But what if they are not?

In 2006, Matthew Salganik and colleagues at Columbia University published a study of an artificial "music market," in which participants could listen to, rate, and download dozens of different tracks. In total there were fourteen thousand participants, whom the researchers secretly split into nine groups. In eight of the groups, participants could see which tracks were popular with their fellow group members. The final group was the control group, in which participants had no idea what others were downloading.

The researchers found that the most popular songs in the control group—a ranking that depended purely on the merits of the songs themselves, and not on what other people were downloading—were not necessarily popular in the eight social groups. In fact, the song rankings in these eight groups varied wildly. Although the "best" songs usually racked up some downloads, mass popularity was not guaranteed. Instead, fame developed in two stages. First, randomness influenced which tracks people happened to pick early on. The popularity of these first downloaded tracks was then amplified by social behavior, with people looking at the rankings and wanting to imitate their peers. "Fame has much less to do with intrinsic quality than we believe it does," Peter Sheridan Dodds, one of the study authors, later wrote, "and much more to do with the characteristics of the people among whom fame spreads."

Mark Roulston and David Hand, statisticians at the hedge fund Winton Capital Management, point out that the randomness of popularity may also influence the ranking of investment funds. "Consider a set of funds with no skill," they wrote in 2013. "Some will produce decent returns simply by chance and these will attract investors, while the poorly performing funds will close and their results may disappear from view. Looking at the results of those surviving funds, you would think that on average they do have some skill."

The line between luck and skill—and between gambling and investing—is rarely as clear as we think. Lotteries should be textbook examples of gambling, but after several weeks of rollovers, they can produce a positive expected payoff: buy up all the combinations of numbers, and you'll make a profit. Sometimes the crossover happens the other way, with investments being more like wagers. Take Premium Bonds, a popular form of investment in the United Kingdom. Rather than receiving a fixed rate of interest as with regular bonds, investors in Premium Bonds are instead entered into a monthly prize draw. The top prize is £1 million, tax-free, and there are several smaller prizes, too. By investing in Premium Bonds, people are in effect gambling the interest they would have otherwise earned. If they instead put their savings in a regular bond, withdrew the interest, and used that money to buy rollover lottery tickets, the expected payoff would not be that different.

If we want to separate luck and skill in a given situation, we must first find a way to measure them. But sometimes an outcome is very sensitive to small changes, with seemingly innocuous decisions completely altering the result. Individual events can have dramatic effects, particularly in sports like soccer and ice hockey where goals are relatively rare. It might be an ambitious pass that sets up a winning shot or a puck that hits the post. How can we distinguish between a hockey victory that is mostly down to talent and one that benefited from lots of lucky breaks?

In 2008, hockey analyst Brian King suggested a way to measure how fortunate a particular NHL player had been. "Let's pretend there was a stat called 'blind luck,'" as he put it. To calculate his statistic, he took the proportion of total shots that a team scored while that player was on the ice and the proportion of opponents' shots that were saved, and then added these two values together. King argued that although creating shooting opportunities involves a lot of skill, there was more luck influencing whether a shot went in or not. Worryingly, when King tested out the statistic on his local NHL team, it showed that the luckiest players were getting contract extensions while the unlucky ones were being dropped.

The statistic, later dubbed "PDO" after King's online moniker, has since been used to assess the fortunes of players—and teams—in other sports, too. In the 2014 soccer World Cup, several top teams failed to make it out of the preliminary group stage. Spain, Italy, Portugal, and England all fell at the first hurdle. Was it because they were lackluster or unlucky? The England team is famously used to misfortune, from disallowed goals to missed penalties. It seems that 2014 was no different: England had the lowest PDO of any team in the tournament, with a score of 0.66.

We might think that teams with a very low PDO are just hapless. Maybe they have a particularly error-prone striker or weak keeper. But teams rarely maintain an unusually low (or high) PDO in the long run. If we analyze more games, a team's PDO will quickly settle down to numbers near the average value of one. It's what Francis Galton called "regression to mediocrity": if a team has a PDO that is noticeably above or below one after a handful of games, it is likely a symbol of luck.

Statistics like PDO can be useful for assessing how lucky teams are, but they aren't necessarily that helpful when placing bets. Gamblers are more interested in making predictions. In other words, they want to find factors that reflect ability rather than luck. But how important is it to actually understand skill?

Take horse races. Predicting events at a racetrack is a messy process. All sorts of factors could influence a horse's performance in a race, from past experience to track conditions. Some of which provide clear hints about the future, while others just muddy the predictions. To pin down which factors are useful, syndicates need to collect reliable, repeated observations about races. Hong Kong was the closest Bill Benter could find to a laboratory setup, with the same horses racing on a regular basis on the same tracks in similar conditions.

Using his statistical model, Benter identified factors that could lead to successful race predictions. He found that some came out as more important than others. In Benter's early analysis, for example, the model said the number of races a horse had previously run was a crucial factor when making predictions. In fact, it was more important than almost any other factor. Maybe the finding isn't all that surprising. We might expect horses that have run more races to be used to the terrain and less intimated by their opponents.

It's easy to think up explanations for observed results. Given a statement that seems intuitive, we can convince ourselves as to why that should be the case, and why we shouldn't be surprised at the result. This can be a problem when making predictions. By creating an explanation, we are assuming that one process has directly caused another. Horses in Hong Kong win *because* they are familiar with the terrain, and they are familiar with it *because* they have run lots of races. But just because two things are apparently related—like probability of winning and number of races run—it doesn't mean that one directly causes the other.

An oft-quoted mantra in the world of statistics is that "correlation does not imply causation." Take the wine budget of Cambridge colleges. It turns out that the amount of money each Cambridge college spent on wine in the 2012–2013 academic year was positively correlated with students' exam results during the same period. The more the colleges spent on wine, the better the results generally were. (King's College, once home to Karl Pearson and Alan Turing, topped the wine list with a spend of £338,559, or about £850 per student.)

Similar curiosities appear in other places, too. Countries that consume lots of chocolate win more Nobel prizes. When ice cream sales rise in New York City, so does the murder rate. Of course, buying ice cream doesn't make us homicidal, just as eating chocolate is unlikely to turn us into Nobel-quality researchers and drinking wine won't make us better at exams.

In each of these cases, there might be a separate underlying factor that could explain the pattern. For Cambridge colleges it could be wealth, which would influence both wine spending and exam results. Or there could be a more complicated set of reasons lurking behind the observations. This is why Bill Benter doesn't try to interpret why some factors appeared to be so important in his horseracing model. The number of races a horse has run might be related to another (hidden) factor that directly influenced performance. Alternatively, there could be an intricate trade-off between races run and other factors—like weight and jockey experience—which Benter could never hope to distill into a neat "A causes B" conclusion. But Benter is happy to sacrifice elegance and explanation if it means having good predictions. It doesn't matter if his factors are counter-intuitive or hard to justify. The model is there to estimate the probability a certain horse will win, not to explain *why* that horse will win.

From hockey to horse racing, sports analysis methods have come a long way in recent years. They have enabled gamblers to study matches in more detail than ever, combining bigger models with better data. As a result, scientific betting has moved far beyond card counting.

ON THE FINAL PAGE of his blackjack book *Beat the Dealer*, Edward Thorp predicted that the following decades would see a whole host of new methods attempting to tame chance. He knew it was hopeless to try to anticipate what they might be. "Most of the possibilities are beyond the reach of our present imagination and dreams," he wrote. "It will be exciting to see them unfold."

Since Thorp made his prediction, the science of betting has indeed evolved. It has brought together new fields of research, spreading far from the felt tables and plastic chips of Las Vegas casinos. Yet the popular image of scientific wagers remains very much in the past. Stories of gambling strategies rarely stray far from the adventures of Thorp or the Eudaemons. Successful betting is viewed as a matter of card counting or watching roulette tables. Tales follow a mathematical path, with decisions reduced to basic probabilities.

But the advantage of simple equations over human ingenuity is not as clear as these stories suggest. In poker, the ability to calculate the probability of getting a particular hand is helpful but by no means a sure route to victory. Gamblers also need to account for their opponents' behavior. When John von Neumann developed game theory to tackle this problem, he found that employing deceptive tactics such as bluffing was actually the optimal thing to do. The gamblers had been right all along, even if they didn't know why.

Sometimes it's necessary to stray from mathematical perfection altogether. As researchers delve further into the science of poker, they are finding situations where game theory comes up short and where traditional gambling traits—reading opponents, exploiting weaknesses, spotting emotion—can help computer players become the best in the world. It is not enough to know just probabilities; successful bots need to combine mathematics and human psychology.

The same is true in sports. Analysts are increasingly trying to capture the individual quirks that make up a team performance.

During the early 2000s, Billy Beane famously used "sabermetrics" to identify underrated players and take the cash-strapped Oakland A's to the Major League Baseball playoffs. The techniques are now appearing in other sports. In the English Premier League, more and more soccer teams are employing statisticians to advise on team performances and potential transfers. When Manchester City won the league in 2014, they had almost a dozen analysts helping to put together tactics.

Sometimes the human element can be the dominant factor, overshadowing the statistics gleaned from available match data. After all, the probability of a goal depends both on the physics of the ball and on the psyche of the player kicking it. Roberto Martinez, manager of Everton soccer club, has suggested that mind-set is as important as performances when assessing potential signings. Managers want to know how a player will settle into a new country or whether he can cope with pressure from a hostile crowd. And, clearly, it is very hard to measure factors like this.

Measurement is often a difficult problem in sports. From the defenders who never make a tackle to the NFL cornerbacks who hardly ever touch the ball, we can't always pin down valuable information. But knowing what we are missing is crucial if we want to fully understand what is happening in a match and what might happen in the future.

When researchers develop a theoretical model of a sport, they are reducing reality to an abstraction. They are choosing to remove detail and concentrate only on key features, much like Pablo Picasso so famously did. When Picasso worked his "Bull" lithographs in the winter of 1945, he started by creating a realistic representation of the animal. "It was a superb, well-rounded bull," said an assistant watching at the time. "I thought to myself that that was that." But Picasso was not finished. After completing his first image, he moved onto a second, and then a third. As Picasso worked on each new picture, the assistant noticed the bull was changing. "It began to diminish, to lose weight," he said. "Picasso was taking away rather than adding to his composition." With each image, Picasso carved further, keeping only the crucial contours, until he reached the eleventh lithograph. Almost every detail had gone, with nothing left but a handful of lines. Yet the shape was still recognizable as a bull. In those few strokes, Picasso had captured the essence of the animal, creating an image that was abstract, but not ambiguous. As Albert Einstein once said of scientific models, it was a case of "everything should be made as simple as possible, but not simpler."

Abstraction is not limited to the worlds of art and science. It is common in other areas of life, too. Take money. Whenever we pay with a credit card, we are replacing physical cash with an abstract representation. The numbers remain the same, but superfluous details—the texture, the color, the smell—have been removed. Maps are another example of abstraction: if a detail is unnecessary, it isn't shown. Weather is abandoned when the focus is on transport and traffic; motorways vanish if we're interested in sun and showers.

Abstractions make a complex world easier to navigate. For most of us, a car accelerator is simply a device that makes the vehicle go faster. We don't care—or need to know—about the chain of events between our foot and the wheels. Likewise, we rarely look at phones as transmitters that convert sound waves to electronic signals; in daily life, they are a series of buttons that produce a conversation.

In fact, it could be argued that our entire notion of randomness is an abstraction. When we say a coin has a 50 percent chance of coming up tails, or that a roulette ball has a 1 in 37 chance of landing on a particular number, we are using an abstraction. In theory we could write down equations for the motion and solve them to predict the trajectory. But because coin flips and roulette spins are so sensitive to initial conditions, it is difficult to this do in reality. So, instead we approximate the process and assume it is unpredictable. We choose to simplify an intricate physical process for the sake of convenience.

In life, we must often choose (either consciously or subconsciously) what abstractions to use. The most extensive abstraction would not omit a single detail. As mathematician Norbert Wiener said, "The best material model of a cat is another, or preferably the same, cat." Capturing the world in such detail is rarely practical, so instead we must strip away certain features. However, the resulting abstraction is our model of reality, influenced by our beliefs and prejudices.

Sometimes abstractions have tried to deliberately influence people's perceptions. In 1947, *Time* magazine published a double-page map of Europe and Asia. Titled the "Communist Contagion," the map's perspective had been altered so the Soviet Union—colored in an ominous red hue—loomed over the rest of the world. The map's creator, a cartographer by the name of R. M. Chapin, continued the theme in subsequent issues. In 1952, a piece called "Europe from Moscow" featured the USSR rising up from the bottom of the image, its borders forming an arrow that pointed toward the West.

Even if the bias is not deliberate, models inevitably depend on their creators' aims (and resources). Recall those different horse racing models: Bolton and Chapman's model had nine factors; Bill Benter used over a hundred. Researchers have to tread a fine line when deciding on an abstraction. Simple models risk omitting crucial features, while complicated models may include unnecessary ones. The trick is to find an abstraction that is detailed enough to be useful, but simple enough to be implementable. In blackjack, for instance, card counters don't need to remember the exact value of each card; they just need enough information to tip the odds in their favor.

Of course, there is always a risk of picking the wrong abstraction, which leaves out a critical detail. Émile Borel once said that, given two gamblers, there is always one thief and one imbecile. This is not just the case when one gambler has much better information than the other. Borel pointed out that in complex situations, two people could have exactly the same information and yet come to different conclusions about the probability of an event. When the pair bet together, Borel said each one would therefore believe "that it is he who is the thief and the other the imbecile."

Poker is a good example of situation in which choice of abstraction is important. There are a huge number of possible moves in poker—far too many to compute—which means that bots have to use abstractions to simplify the game. Tuomas Sandholm has pointed out that this can cause problems. For instance, your bot might only think in terms of certain bet sizes, to avoid having to analyze every possible wager. Over time, however, the bot's view of reality will not match the true situation. "Your belief as to how much money is in the pot is no longer accurate," Sandholm said. This can leave you vulnerable to an opponent using a better abstraction, which is closer to reality.

The problem doesn't just appear in poker. The entire casino industry is built on the assumption that the games are random. Casinos treat roulette spins and blackjack shuffles as unpredictable and rely on customers sharing that view. But believing an abstraction doesn't make it correct. And when someone comes along with a better model of reality—someone like Edward Thorp or Doyne Farmer—that person can profit from the casinos' oversimplification.

Thorp and Farmer were both physics students when they began their work on casino games. In subsequent decades, other students and academics have followed their lead. Some have targeted casinos, while others have focused on sports and horse racing. Which raises the question: Why is betting so popular among scientists?

IN JANUARY 1979, A group of undergraduates at Massachusetts Institute of Technology set up an extracurricular course called "How to Gamble If You Must." It was part of the university's four-week-long independent activities period (IAP), which encouraged students to take new classes and broaden their interests. During the gambling course, participants learned about Thorp's blackjack strategy and how to count cards. Soon some of the players had decided to try out the tactics for real; first in Atlantic City, and then in Vegas.

Although the players had started with Thorp's methods, they needed a new approach if they were going to be successful. As Thorp had discovered, it was difficult to get away with solo card counting. Players have to raise their bets when the count is in their favor, which means they are likely to attract the attention of casino security. The MIT students therefore worked as a team. Some players would be spotters, whose job it was to bet the minimum stake while keeping track of the count. When the deck was sufficiently in their favor, the spotters would signal to another group—the "big players"—who would come and throw lots of money at the table. To help conceal their roles from security, the team exploited common casino stereotypes. Smart female students would put on low-cut tops and pretend to be dumb gamblers, all the while keeping count of the cards. Students with an Asian or Middle Eastern background would play the role of a rich foreigner, happy to spend their parents' money.

Although its members changed over time, the MIT team continued to take on the casinos for many years. The contrast with life in Massachusetts could not have been larger. Instead of dorm rooms and Boston rain, there were hotel suites, sunny skies, and huge profits. During the Fourth of July weekend in 1995, the team was so successful that when they met by the pool at the end of the trip, one of them was carrying a gym bag holding almost a million dollars in cash. Another time, one of the team left a paper bag containing $125,000 in a classroom at MIT. When they returned, the bag was gone. They later discovered the janitor had stored it in his locker; they only got the money back after six months of investigations by the FBI and Drug Enforcement Administration.

The MIT blackjack team has become part of gambling legend. Journalist Ben Mezrich told their story in the best-selling book *Bringing Down the House*, and the events later inspired the film *21*. Unfortunately, for modern students, however, the exploits of the MIT team have become history in more ways than one. Casinos have introduced lots more countermeasures in recent years, which means teams would struggle to reproduce the sort of success seen in the 1980s and 1990s. In fact, according to professional gambler Richard Munchkin, hardly anyone focuses exclusively on blackjack anymore. "I know very few people—people you could count on one hand—who are making a living only by counting cards," he said.

Yet the science of gambling still features at MIT. In 2012, PhD student Will Ma set up a new course as part of the independent activities period. Its official title was "15.S50," but everybody knew it as the MIT poker class. Ma, who was studying operations research, had played a lot of poker—and won a lot of money—during his undergraduate days in Canada. When he arrived at MIT, word of his success got out, and several people started asking him questions about poker. One of them was his head of department, Dimitris Bertsimas, who also had an interest in the game. Bertsimas helped Ma put together a class to teach the theory and tactics needed to win. It was a legitimate MIT class; if students passed, they could get degree credit.

The course attracted a lot of attention. In fact, so many people turned up for the first class that they had to move rooms. "It was probably one of the most popular classes during IAP," Ma said. Attendees ranged from undergraduate business students to PhD mathematicians. Ma's class also caught the eye of the online poker community. Many incorrectly believed that students were going to use their expertise to build poker software. "Through word of mouth, it somehow got twisted," Ma said. "They thought it was going to lead to a huge poker bot system with a ton of bots written by MIT students taking all the money."

As well as distancing himself from bots, Ma also had to be careful to avoid his course on poker being misinterpreted by the university. "It can be seen as gambling," he said, "and you're not supposed to teach gambling at MIT." He therefore used play money to demonstrate strategies. "I had to make sure I wasn't taking people's real money."

Ma didn't have enough time to cover every aspect of poker, so instead he tried to focus on topics that would provide the biggest benefits. "I tried to go through the steepest part of the learning curve," he said. He explained why players shouldn't be afraid to go in at the start of a round and the dangers of getting bored with folding and instead playing too many hands. Many of the lessons would be useful in other situations, too. "I tried to put it in the perspective of real life," Ma said. The poker class covered the importance of making confident moves and of not letting mistakes affect performance. Students learned how to read opponents and how to manage the image they conveyed during games. In doing so, they started to discover what luck and skill really looked like. "I think one of the things poker teaches you very well is that you can often make a good decision but not get a good result," Ma said, "or make a bad decision and get a good result."

COURSES TEACHING THE SCIENCE of gambling have cropped up in other institutions, too, from York University in Ontario to Emory University in Georgia. In these classes, students study lotteries, roulette, card shuffling, and horse races. They learn statistics and strategy, analyzing risks and weighing options. Yet, as Ma found, people can be hostile to the concept of betting in universities. Indeed, many people are against the idea of wagers in any context.

When people say they dislike betting, what they usually mean is that they dislike the betting industry. Although the two are related, they are by no means synonymous. Even if we never gambled in casinos or visited bookmakers, betting would still permeate our lives. Luck—good and bad—looms over our careers and relationships. We have to deal with hidden information and negotiate in the face of uncertainty. Risks must be balanced with rewards; optimism must be weighed against probability.

The science of gambling isn't just useful for gamblers. Studying betting is a natural way to explore the notion of luck and can therefore be a good way to hone scientific skills. Although Ruth Bolton and Randall Chapman's paper on horse racing predictions gave rise to a multi-billion-dollar betting industry, it was the only article Bolton wrote on the topic. She has spent the rest of her career working on other problems. Most of them revolve around marketing, from the effects of different pricing strategies to how businesses can manage customer relationships. Bolton admits that the horse racing paper could therefore seem like a bit of an outlier on her CV; at first glance, it doesn't really fit in with her other research. But the methods in that early racetrack study, which involved developing models and assessing potential outcomes, would go on to shape the rest of her work. "That way of thinking about the world stayed with me," she said.

Probability theory, which Bolton used to analyze horse races, is one of the most valuable analytical tools ever created. It gives us the ability to judge the likelihood of events and assess the reliability of information. As a result, it is a vital component of modern scientific research, from DNA sequencing to particle physics. Yet the science of probability emerged not in libraries or lecture theaters but among the cards and dice of bars and game rooms. For eighteenth-century mathematician Pierre Simon Laplace, it was a strange contrast. "It is remarkable that a science which began with the consideration of games of chance should have become the most important object of human knowledge."

Cards and casinos since have inspired many other scientific ideas. We have seen how roulette helped Henri Poincaré develop the early ideas of chaos theory and allowed Karl Pearson to test his new statistical techniques. We also met Stanislaw Ulam, whose card games led to the Monte Carlo method, now used in everything from 3D computer graphics to the analysis of disease outbreaks. And we have seen how game theory emerged from John von Neumann's analysis of poker.

The relationship between science and betting continues to thrive today. As ever, the ideas are flowing in both directions: gambling is inspiring new research, and scientific developments are providing new insights into betting. Researchers are using poker to study artificial intelligence, creating computers that can bluff and learn and surprise just like humans. Every year, these champion bots are coming up with new tactics that humans never knew about, or would never dare try. Meanwhile, high-speed algorithms are helping companies make bets and trades automatically, creating a complex ecosystem of interactions that has prompted new avenues of research. Sports analysts, armed with better data and faster computers, are no longer just predicting team results; they are picking apart the roles of individual players, measuring the contribution of chance and skill. From poker to betting exchanges, researchers are developing a deeper understanding of human behavior and decision making, and in turn coming up with more effective gambling strategies.

THE POPULAR IMAGE OF a scientific betting strategy is one of a mathematical magic trick. To get rich, all you need is a simple formula or a few basic rules. But, much like a magic trick, the simplicity of the performance is an illusion, concealing a mountain of preparation and practice.

As we've seen, almost any game can be beaten. But profits rarely come from lucky numbers or "foolproof" systems. Successful wagers take patience and ingenuity. They require creators who choose to ignore dogma and follow their curiosity. It might be a student like James Harvey, who wondered which lottery was the best deal and orchestrated thousands of ticket purchases to take advantage of the loophole he found. Or a physicist like Edward Thorp, rolling marbles on his kitchen floor to understand where a roulette ball would stop. It might take a business specialist like Ruth Bolton, crunching through horse racing data to find out what makes a winner. Or statisticians such as Mark Dixon and Stuart Coles, reading an undergraduate exam question about soccer prediction and wondering how the methods could be improved.

From the casinos of Monte Carlo to the racetracks of Hong Kong, the story of the perfect bet is a scientific one. Where once there were rules of thumb and old wives' tales, there are now theories guided by experiment. The reign of superstition has waned, usurped by rigor and research. Bill Benter, who made his fortune betting on blackjack and horse racing, has no doubts about who deserves the credit for the transition. "It wasn't as though streetwise Las Vegas gamblers figured out a system," he said. "Success came when an outsider, armed with academic knowledge and new techniques, came in and shone light where there had been none before."

ACKNOWLEDGMENTS

FIRST, THANKS MUST GO to my agent Peter Tallack. From proposal to publisher, his guidance over the past three years has been invaluable. I would also like to thank my editors—TJ Kelleher and Quynh Do at Basic Books, and Nick Sheerin at Profile—for taking a gamble on me, and for helping me shape the science into a story.

My parents continue to provide crucial suggestions and discussions on my writing, and for this I am eternally grateful. Thanks also to Clare Fraser, Rachel Humby, and Graham Wheeler for many useful comments on early drafts. And, of course, to Emily Conway, who has been there for me throughout with wise words and wine.

Finally, I am indebted to everyone who took the time to share their insights and experiences: Bill Benter, Ruth Bolton, Neil Burch, Stuart Coles, Rob Esteva, Doyne Farmer, David Hastie, Michael Johanson, Michael Kent, Will Ma, Matt Mazur, Richard Munchkin, Brendan Poots, Tuomas Sandholm, Jonathan Schaeffer, Michael Small, and Will Wilde. Many of these individuals have shaped entire industries with their scientific curiosity. It will be fascinating to see what comes next.

NOTES

Introduction

ix **In June 2009, a British newspaper:** Ward, Simon. "A Sacked 22-Year-Old Trainee City Trader Today Reveals How He Won a Staggering £20 Million in a Year . . . Betting on the Horses." *News of the World,* June 26, 2009.

ix **He had a chauffeur-driven Mercedes:** Duell, Mark. "'King of Betfair' Who Lived Lavish Lifestyle in Top Hotels with Chauffeur-Driven Mercedes and Clothes from Harrods after Conning Family Friends Out of £400,000 Is Jailed." *Daily Mail* Online, May 28, 2013. http://www.dailymail.co.uk/news/article-2332115/King-Betfair-stayed-hotels-splashed-chauffeur-conning-family-friends-jailed.html.

ix **The profitable bets that Short claimed:** Wood, Greg. "Short Story on Betfair System Is Pure Fiction." *Guardian Sportblog* (blog), June 29, 2009. http://www.theguardian.com/sport/blog/2009/jun/30/greg-wood-betfair-notw-story.

ix **Having persuaded investors to pour hundreds of thousands:** Duell, Mark. "Gambler, 26, Who Called Himself the 'Betfair King' Conned Friends Out of £600,000 with Betting Scam to Pay for Designer

Clothes." *Daily Mail* Online, April 23, 2013. http://www.dailymail.co.uk/news/article-2313618/Gambler-called-Betfair-king-conned friends-600–000-bogus-betting-scam.html.

x **the case went to trial:** "Criminal Sentence—Elliott Sebastian Short—Court: Southwark." TheLawPages.com, May 28, 2013. http://www.thelawpages.com/court-cases/Elliott-Sebastian-Short-11209–1.law.

x **As its reputation spread:** Ethier, Stewart. *The Doctrine of Chances: Probabilistic Aspects of Gambling* (New York: Springer, 2010), 115.

xi **"The martingale is as elusive":** Dumas, Alexandre. *One Thousand and One Ghosts* (London: Hesperus Classics, 2004).

xi **Having frittered away his inheritance:** O'Connor, J. J., and E. F. Robertson. "Girolamo Cardano." June 1998. http://www-history.mcs.st-andrews.ac.uk/Biographies/Cardan.html.

xi **nobody knew precisely what a "fair" wager:** O'Connor and Robertson, "Girolamo Cardano."

xi **deriving "Cardano's formula":** Gorroochurn, Prakash. "Some Laws and Problems of Classical Probability and How Cardano Anticipated Them." *Chance Magazine* 25, no. 4 (2012): 13–20.

xii **"When I observed that the cards were marked":** Cardan, Jerome. *Book of My Life* (New York: Dutton, 1930).

xii **At the request of a group of Italian nobles:** Ore, Oystein. "Pascal and the Invention of Probability Theory." *American Mathematical Monthly* 67, no. 5 (May 1960): 409–419.

xii **Astronomer Johannes Kepler also took time:** Epstein, Richard. *The Theory of Gambling and Statistical Logic* (Waltham, MA: Academic Press, 2013).

xii **The science of chance blossomed:** Ore, "Pascal and the Invention."

xii **he was more likely to get a six:** It's easiest to start by calculating the probability of *not* getting a six in four rolls, which is $(5/6)^4$. It therefore follows that the probability of getting at least one six is $1–(5/6)^4 = 51.8$ percent. By the same logic, the probability of getting a double six over twenty-four throws of two dice is $1–(35/36)^4 = 49.1$ percent.

xii **"Gamblers can rightly claim":** Epstein, Richard. *The Theory of Gambling and Statistical Logic* (Waltham, MA: Academic Press, 2013).

xiii **Daniel Bernoulli wondered why people:** Bassett, Gilbert, Jr. "The St. Petersburg Paradox and Bounded Utility." *History of Political Economy* 19, no. 4 (1987): 517–523.

xiii **"The mathematicians estimate money":** Castelvecchi, Davide. "Economic Thinking." *Scientific American* 301, no. 82 (September 2009). doi:10.1038/scientificamerican0909–82b.

xiv **When Feynman tried the game:** Feynman, Richard. *Surely You're Joking, Mr. Feynman!* (New York: W. W. Norton, 2010).

Chapter 1

1 **It's called the Ritz Club:** Ritz Club brochure.

1 **One evening in March 2004:** Chittenden, Maurice. "Laser-Sharp Gamblers Who Stung Ritz Can Keep £1.3m." *Times* (London), December 5, 2004.

1 **The group weren't like the other high rollers:** Beasley-Murray, Ben. "Special Report: Wheels of Justice." *PokerPlayer*, January 1 2005. http://www.pokerplayer365.com/uncategorized-drafts/wheels-of-justice/.

2 **This time their winnings:** "'Laser Scam' Gamblers to Keep £1m." BBC News Online, December 5, 2004. http://news.bbc.co.uk/2/hi/uk/4069629.stm.

2 **What they saw was enough:** Chittenden, "Laser-Sharp Gamblers."

2 **It was one of his many interests:** Mazliak, Laurent. "Poincaré's Odds." *Séminaire Poincaré* XVI (2012): 999–1037.

2 **As Poincaré saw it:** Poincaré, Henri. *Science and Hypothesis* (New York: Walter Scott Publishing, 1905). (French edition published in 1902)

3 **suppose we drop a can of paint:** According to Scott Patterson, Edward Thorp once did this at a pool in Long Beach, California (with red dye rather than paint). The incident made the local paper. Source: Patterson, Scott. *The Quants* (New York: Crown, 2010).

3 **Instead, we can simply watch:** Poincaré, Henri. *Science and Method* (London: Nelson, 1914). (French edition published in 1908)

3 **Taking time off from their studies:** Ethier, Stuart. "Testing for Favorable Numbers on a Roulette Wheel." *Journal of the American Statistical Association* 77, no. 379 (September 1982): 660–665.

4 **Pearson got a colleague to flip a penny:** Pearson, K. "The Scientific Aspect of Monte Carlo Roulette." *Fortnightly Review*, February 1894.

4 **we have "no absolute knowledge of natural phenomena:** Pearson, K. *The Ethic of Freethought and Other Addresses and Essays* (London: T. Fisher Unwin, 1888).

5 **He was particularly keen on German culture:** Magnello, M. E. "Karl Pearson and the Origins of Modern Statistics: An Elastician Becomes a Statistician." *Rutherford Journal.* http://www.rutherfordjournal.org/article010107.html.

5 **the newspaper *Le Monaco*:** Pearson, "Scientific Aspect of Monte Carlo Roulette."

6 **Gamblers crowded around the table:** Huff, Darrell, and Irving Geis. *How to Take a Chance* (London: W. W. Norton, 1959), 28–29.

6 **Monte Carlo roulette confounds his theories:** Pearson, "Scientific Aspect of Monte Carlo Roulette."

7 **the reporters had decided it was easier:** MacLean, L. C., E. O. Thorp, and W. T. Ziemba, eds. *The Kelly Capital Growth Investment Criterion: Theory and Practice* (Singapore: World Scientific, 2011).

7 **Reports of their final profits differ:** Maugh, Thomas H. "Roy Walford, 79; Eccentric UCLA Scientist Touted Food Restriction." *Los Angeles Times*, May 1, 2004. http://articles.latimes.com/2004/may/01/local/me-walford1.

7 **Many have told the tale:** Ethier, "Testing for Favorable Numbers."

7 **When Wilson published his data:** Ethier, "Testing for Favorable Numbers."

9 **Poincaré had outlined the "butterfly effect:** Gleick, James. *Chaos: Making a New Science* (New York: Open Road, 2011).

9 **The Zodiac may be regarded:** Poincaré, *Science and Method.*

10 **Blaise Pascal invented roulette:** Bass, Thomas. *The Newtonian Casino* (London: Penguin, 1990).

10 **The orbiting roulette ball:** The majority of details and quotes in this section are taken from Thorp, Edward. "The Invention of the First Wearable Computer." *Proceedings of the 2nd IEEE International Symposium on Wearable Computers* (1998), 4.

13 **participants were asked to help:** Milgram, Stanley. "The Small-World Problem." *Psychology Today* 1, no. 1 (May 1967): 61–67.

13 **an average of 3.74 degrees of separation:** Backstrom, Lars, Paolo Boldi, Marco Rosa, Johan Ugander, and Sebastiano Vignal. "Four Degrees of Separation" (Cornell University Library, January 2012). http://arxiv.org/abs/1111.4570.

14 **Another attendee was a young physicist:** Gleick, "Chaos."

14 **By taking measurements:** Bass, *Newtonian Casino.*

14 **When a new paper on roulette appeared:** Small, Michael, and Chi

Kong Tse. "Predicting the Outcome of Roulette." *Chaos* 22, no. 3 (2012): 033150. doi:10.1063/1.4753920.

15 **For his PhD, he'd analyzed:** Quotes and additional details come from an interview with Michael Small in 2013.

18 **He was sailing in Florida:** Author interview with Doyne Farmer, October 2013.

19 **He'd found that air resistance:** Slezak, Michael. "Roulette Beater Spills Physics behind Victory." *New Scientist*, no. 2864 (May 12, 2012). https://www.newscientist.com/article/mg21428644-500-roulette-beater-spills-physics-behind-victory/. Additional details from author interview with Doyne Farmer, October 2013.

19 **During their casino trips:** Bass, *Newtonian Casino*.

19 **To predict exactly where the cue ball will travel:** Crutchfield, James P., J. Doyne Farmer, Norman H. Packard, and Robert S. Shaw. "Chaos." *Scientific American* 254, no. 12 (December 1986): 46–57.

20 **when journalist Ben Beasley-Murray talked:** Details about subsequent investigations are from Beasley-Murray, Ben. "Special Report: Wheels of Justice." *PokerPlayer*, January 1, 2005. http://www.pokerplayer365.com/uncategorized-drafts/wheels-of-justice/.

20 **According to ex-Eudaemon Norman Packard:** McKee, Maggie. "Alleged High-Tech Roulette Scam 'Easy to Set Up.'" *New Scientist*, March 2004.

21 **When Hibbs and Walford passed $5,000:** Ethier, "Testing for Favorable Numbers."

Chapter 2

23 **Of the colleges of the University of Cambridge:** Gonville and Caius. "History." http://www.cai.cam.ac.uk/history.

23 **its unique stained glass windows:** Author experience.

23 **Fisher spent three years studying at Cambridge:** O'Connor, J. J., and E. F. Robertson. "Sir Ronald Aylmer Fisher." JOC/EFR, October 2003. http://www-history.mcs.st-and.ac.uk/Mathematicians/Fisher.html.

24 **"To consult the statistician after an experiment . . .":** Fisher, Ronald. "Presidential Address to the First Indian Statistical Congress." *Sankhya* 4 (1938):14–17.

25 **The Great Wall of China was financed:** Campbell, Alex. "National

Lottery: Why Do People Still Play?" BBC News Online, October 2013. http://www.bbc.com/news/uk-24383871.

25 **proceeds from a lottery organized in 1753:** Wilson, David. "The British Museum: 250 Years On." *History Today* 52 (2002): 10.

25 **many of the Ivy League universities were built:** Lehrer, Jonah. "Cracking the Scratch Lottery Code." *Wired*, January 31, 2011. http://www.wired.com/2011/01/ff_lottery/.

26 **a quarter of the National Lottery's revenues:** Bowers, Simon. "Lottery Scratchcards Fuel Camelot Sales Boom." *Guardian*, November 18, 2011. http://www.theguardian.com/uk/2011/nov/18/national-lottery-scratchcard-sales-boom.

26 **American state lotteries earn tens of billions:** Scratchcards.org. "The Lottery Industry." http://www.scratchcards.org/featured/57121/the-lottery-industry.

26 **To quote statistician William Gossett:** Ziliak, Stephen. "Balanced Versus Randomized Field Experiments in Economics: Why W. S. Gosset aka 'Student' Matters." *Review of Behavioral Economics* 1, no. 1–2 (2014): 167–208. http://dx.doi.org/10.1561/105.00000008.

26 **how the lottery keeps track:** Lehrer, "Cracking the Scratch Lottery Code."

26 **he'd known Bill Tutte:** Yang, Jennifer. "Toronto Man Cracked the Code to Scratch-Lottery Tickets." *Toronto Star*, February 4, 2011. http://www.thestar.com/news/gta/2011/02/04/toronto_man_cracked_the_code_to_scratchlottery_tickets.html.

26 **British mathematician who had broken the Nazi Lorenz cipher:** William Tutte obituary. *Kitchener-Waterloo Record*, May 2002.

27 **a computer version of tic-tac-toe:** Yang, "Toronto Man Cracked the Code."

27 **By the time she was arrested:** George, Patrick. "Woman Crashes Car into Convenience Store to Steal 1,500 Lotto Tickets." MSN Online, May 13, 2013. http://jalopnik.com/woman-crashes-car-into-convenience-store-to-steal-1–500–504608879.

27 **"We need to talk":** Yang, "Toronto Man Cracked the Code."

28 **Joan Ginther had won four jackpots:** Rich, Nathanial. "The Luckiest Woman on Earth." *Harper's Magazine*, August 2011.

28 **wanted to call the dorm "Random House":** Roller, Dean. "Publisher's Objections Force New Dorm Name." *The Tech*, January 1968.

http://web.mit.edu/~random-hall/www/History/publisher-objections.shtml.

28 **selling the naming rights on eBay:** eBay. "eBay Item # 1700894687 Name a Floor at MITs Random Hall." http://web.mit.edu/ninadm/www/ebay.html.

28 **The hall even has its own:** Dowling, Claudia. "MIT Nerds." *Discover Magazine*, June 2005.

28 **he became interested in lotteries:** Details of the Powerball syndicate activities come from Sullivan, Gregory. "Letter to State Treasurer Steven Grossman." July 2012. http://www.mass.gov/ig/publications/reports-and-recommendations/2012/lottery-cash-winfall-letter-july-2012.pdf.

32 **"It took us about a year to ramp up to it":** Sullivan, Letter to State Treasurer Steven Grossman.

32 **The effort paid off:** Estes, Andrea. "A Game of Chance Became Anything But." *Boston Globe*, October 16, 2011. http://www.boston.com/news/local/massachusetts/articles/2011/10/16/a_game_of_chance_became_anything_but/.

32 **the *Boston Globe* had published a story:** Estes, "Game of Chance."

33 **Selbee claims to have won around $8 million:** Wile, Rob. "Retiree from Rural Michigan Tells Us the Moment He Figured Out How to Beat the State's Lottery." *Business Insider*, August 1, 2012. http://www.businessinsider.com/a-retiree-from-rural-michigan-tells-us-the-moment-he-figured-out-how-to-beat-the-states-lottery-2012–8.

CHAPTER 3

35 **"Professional card counters are prohibited":** Yafa, Stephen. "In the Cards." *The Rotarian*, November 2011.

35 **Thorp has been called the father of card counting:** Many sources have made this reference. One prominent example is the publisher blurb for: Thorp, Edward. *Beat the Dealer* (New York: Random House, 1962).

36 **When one of the men suggested a game of blackjack:** Kahn, Joseph P. "Legendary Blackjack Analysts Alive but Still Widely Unknown." *The Tech*, February 2008. http://tech.mit.edu/V128/N6/blackjack.html.

36 **the dealer had a 40 percent chance:** Baldwin, Roger, Wilbert E. Cantey, Herbert Maisel, and James P. McDermott. "The Optimum

Strategy in Blackjack." *Journal of the American Statistical Association* 51, no. 275 (1956): 429–439.

37 **Intrigued by Baldwin's idea:** Haney, Jeff. "They Invented Basic Strategy." *Las Vegas Sun News*, January 4, 2008.

37 **"In statistical terms":** Kahn, "Legendary Blackjack Analysts Alive."

37 **The four men published their findings:** Baldwin et al., "The Optimum Strategy in Blackjack." The four soldiers also later published a book for nonstatisticians, entitled *Playing Blackjack*. According to McDermott, it made a total of $28. (Source: Kahn, "Legendary Blackjack Analysts Alive.")

37 **It was meant to be a relaxing holiday:** Thorp, Edward. *Beat the Dealer* (New York: Random House, 1962).

38 **Thorp gradually turned the research:** Kahn, "Legendary Blackjack Analysts Alive."

38 **He saw it more as an academic obligation:** Towle, Margaret. "Interview with Edward O. Thorp." *Journal of Investment Consulting* 12, no. 1 (2011): 5–14.

39 **"It showed that nothing was invulnerable":** Author interview with Bill Benter, July 2013.

39 **Switching his university campus in Cleveland:** Yafa, Stephen. "In the Cards." *The Rotarian*, November 2011.

39 **The decision was to prove extremely lucrative:** Ibid.

39 **his firm was commissioned by the Australian government:** Dougherty, Tim. "Horse Sense." *Contingencies*, June 2009.

40 **"It's easy to learn how to count cards":** Author interview with Richard Munchkin, August 2013.

40 **To evade security:** Thorp, *Beat the Dealer*.

40 **Most mathematicians in the early twentieth century:** Mazliak, Laurent. "Poincaré's Odds." *Séminaire Poincaré* XVI (2002): 999–1037.

41 **His research is still used today:** Saloff-Coste, Laurent. "Random Walks on Finite Groups." In *Probability on Discrete Structures*, ed. Harry Kesten (New York: Springer Science & Business, 2004).

41 **To perform the shuffle:** Blood, Johnny Blood. "A Riffle Shuffle Being Performed during a Game of Poker at a Bar Near Madison, Wisconsin,. November 2005–April 2006." Source: Flickr. Image licensed under CC-BY-SA 2.0.

41 **Figure 3.1. A dovetail shuffle:** Reproduced under the CC BY-SA 2.0 license. https://www.flickr.com/photos/latitudes/66424863.

42 **For a fifty-two-card deck:** Bayer, D. B., and P. Diaconis. "Trailing the Dovetail Shuffle to Its Lair." *Annals of Applied Probability* 2, no. 2 (1992): 294–313.

42 **Benter found that casinos:** Author interview with Bill Benter, July 2013.

42 **They would enter information:** Schnell-Davis, D. W. "High-Tech Casino Advantage Play: Legislative Approaches to the Threat of Predictive Devices." *UNLV Gaming Law Journal* 3, no. 2 (2012).

42 **Unfortunately for gamblers:** Author interview with Richard Munchkin, August 2013.

43 **"Once you become well known":** Author interview with Bill Benter, July 2013.

43 **Cheers rise above the sound:** Author experience.

43 **an average of $145 million:** Lee, Simon. "Hong Kong Horse Bets Hit Record as Races Draw Young Punters." *BusinessWeek*, July 11, 2013. http://www.bloomberg.com/news/articles/2013–07–11/hong-kong-horse-bets-hit-record-as-races-draw-young-punters.

43 **the Kentucky Derby set a new American record:** "Record-Breaking Day Across-the-Board for Kentucky Derby 138." Kentucky Derby, May 6, 2012. http://www.kentuckyderby.com/news/2012/05/05/kentucky-derby-138-establishes-across-board-records.

43 **The Jockey Club is a nonprofit organization:** Rarick, Gina. "Horse Racing: Hong Kong Polishes a Good Name Worth Gold." *New York Times*, December 11, 2004. http://www.nytimes.com/2004/12/11/sports/11iht-horse_ed3_.html?_r=0.

44 **Undeterred, Julius tweaked the mechanism:** Doran, Bob. "The First Automatic Totalisator." *Rutherford Journal*. http://rutherfordjournal.org/article020109.html.

45 **sports bettors need a strategy:** Benter, William. "Computer Based Horse Race Handicapping and Wagering Systems: A Report." In *Efficiency of Racetrack Betting Markets*, ed. D. B. Hausch, V. S. Y. Lo, and W. T. Ziemba (London: Academic Press, 1994), 511–526.

45 **Unfortunately, it was a year:** Dougherty, "Horse Sense."

46 **"Searching for Positive Returns at the Track":** Bolton, R. N., and R. G. Chapman. "Searching for Positive Returns at the Track: A Multinomial Logit Model for Handicapping Horse Races." *Management Science* 32, no. 8 (1986).

46 **"It was the paper that launched":** Author interview with Bill Benter, July 2013.

46 **Karl Pearson met a gentleman:** Magnello, M. Eileen. "Karl Pearson and the Origins of Modern Statistics: An Elastician Becomes a Statistician." *Rutherford Journal*. http://www.rutherfordjournal.org/article010107.html.

46 **"He never waited to see":** Pearson, Karl. *The Life, Letters and Labours of Francis Galton* (Cambridge: Cambridge University Press, 2011).

47 **seven of Galton's friends received sweet pea seeds:** Galton, Francis. "Towards Mediocrity in Hereditary Stature." *Journal of the Anthropological Institute of Great Britain and Ireland* 15 (1986): 246–263.

48 **Galton was so impressed:** Galton, Francis. "A Diagram of Heredity." *Nature* 57 (1898): 293. http://www.esp.org/foundations/genetics/classical/fg-98.pdf.

48 **Both viewed regression to the mediocre:** Pearson, *Life, Letters and Labours*.

49 **In Pearson's view, a nation could be improved:** Pearson, Karl. *National Life from the Standpoint of Science*, 2nd ed. (Cambridge: Cambridge University Press, 1919).

49 **he also claimed that laws:** Pearson, Karl. "The Problem of Practical Eugenics" (Galton Eugenics Laboratory Lecture Series No. 5. Dulau & Co., 1909).

49 **"When I was a toddler":** Author interview with Ruth Bolton, February 2014.

52 **Some gamblers might try to think up:** Author interview with Bill Benter, July 2013.

52 **The task fell to William Gossett:** Ziliak, Stephen. "Guinnessometrics: The Economic Foundation of 'Student's' *t*." *Journal of Economic Perspectives* 22, no. 4 (2008): 199–216.

53 **The pilots, busy with other tasks:** Emanuel, Kerry. "Edward Norton Lorenz 1917–2008." *National Academy of Sciences*, 2011. ftp://texmex.mit.edu/pub/emanuel/PAPERS/Lorenz_Edward.pdf.

55 **In 1994, Benter published a paper:** Benter, "Computer Based Horse Race Handicapping."

55 **Researching a company to its core:** Investopedia.com. "Technical Analysis." http://www.investopedia.com/terms/t/technicalanalysis.asp.

57 **"People had so little faith in the system":** Dougherty, "Horse Sense."

57 **Disagreements meant the partnership ended:** Dougherty, "Horse Sense."

58 **As a result, both horses take the same time:** Author interview with Bill Benter, July 2013.

58 **Finding his brain severely inflamed:** Grimberg, Sharon (producer/writer) and Rick Groleau (producer/writer). "Race for the Superbomb," directed by Thomas Ott, aired for The American Experience series (PBS Video, 1999).

59 **It wasn't his first choice:** Rota, Gian-Carlo. "The Lost Café." *Los Alamos Science*, 1987.

59 **When he finally got the answer:** Rota, "The Lost Café."

60 **"Suddenly, I knew where":** Lounsberry, Alyse. "A-Bomb Cloaked in Mystery." *Ocala Star-Banner*, December 4, 1978, 13.

60 **"This train is linear, isn't it?":** Halmos, Paul. "The Legend of John von Neumann." *American Mathematical Monthly* 8 (1973): 382–394.

61 **"Any one who considers arithmetical methods":** Von Neumann, John. "Various Techniques Used in Connection with Random Digits." *Journal of Research of the National Bureau of Standards*, Appl. Math. Series, 1951. Quoted in Herman Heine Goldstine, *The Computer from Pascal to von Neumann* (Princeton, NJ: Princeton University Press, 2008).

62 **"Andrei Neistovy": Andrei the angry:** Mazliak, Laurent. "From Markov to Doeblin: Events in Chain" (talk given at RMR-2010, Rouen, France, June 1, 2010). http://www.proba.jussieu.fr/~mazliak/Markov_Rouen.pdf.

62 **a local prison psychologist turned up:** Described in Diaconis, P. "The Markov Chain Monte Carlo Revolution." *Bulletin of the American Mathematical Society* 46 (2009): 179–205.

64 **Metropolis and his colleagues realized:** Metropolis, Nicholas, Arianna W. Rosenbluth, Marshall N. Rosenbluth, Augusta H. Teller, and Edward Teller. "Equation of State Calculations by Fast Computing Machines." *Journal of Chemical Physics* 21 (1953): 1087. http://dx.doi.org/10.1063/1.1699114.

64 **Markov chain Monte Carlo has helped syndicates:** Author interview with Bill Benter, July 2013. Key citation: Gu, Ming Gao, and Fan Hui Kong. "A Stochastic Approximation Algorithm with Markov

Chain Monte-Carlo Method for Incomplete Data Estimation Problems." *Proceedings of the National Academy of Science USA* 95 (1998): 7270–7274.

65 **The formula is named after John Kelly:** Poundstone, W. *Fortune's Formula: The Untold Story of the Scientific Betting System That Beat the Casinos and Wall Street* (New York: Hill and Wang, 2006).

66 **Consistently overestimate by twofold:** Chapman, S. "The Kelly Criterion for Spread Bets." *IMA Journal of Applied Mathematics* 72 (2007): 43–51.

66 **This reduces the risk:** Benter, "Computer Based Horse Race Handicapping."

67 **"The late money tends to be smart money":** Author interview with Bill Benter, July 2013.

67 **The techniques are now so effective:** Dougherty, "Horse Sense."

68 **Benter and Woods chose Hong Kong:** Author interview with Bill Benter, July 2013.

68 **teams using computer predictions bet around $2 billion:** Description of current events comes from Jagow, Scott. "I, Robot: The Future of Horse Wagering?" *Paulick Report*, 2013. http://www.paulickreport.com/news/ray-s-paddock/i-robot-the-future-of-horse-race-wagering/.

68 **Like Swedish harness racing:** Author interview with Bill Benter, July 2013.

69 **"We would joke that we could do it":** Author interview with Ruth Bolton, February 2014.

Chapter 4

71 **When a new blackjack system hit Britain:** Author experience.

71 **The strategy emerged as a result:** House of Commons Culture, Media and Sport Committee. "The Gambling Act 2005: A Bet Worth Taking?" HC 421, 2012.

72 **The first conviction for bonus abuse:** "Man Jailed for 'Bonus Abuse.'" Metropolitan Police Online, April 2012. http://content.met.police.uk/News/Man-jailed-for-Bonus-Abuse/1400007796996/1257246745756.

72 **casinos rarely understand how much of an advantage:** Interview with Richard Munchkin, August 2013.

73 **The events have more in common:** Author experience.

74 **it is the science of the very unlikely:** De Haan, L., and A. Ferreira. Preface. In *Extreme Value Theory: An Introduction* (New York: Springer Science & Business Media, 2007).

74 **Coles's research spanned everything:** For example: Coles, Stuart, and Jonathan Tawn. "Bayesian Modelling of Extreme Surges on the UK East Coast." *Philosophical Transactions* 363, no. 1831 (2005): 1387–1406; and Coles, Stuart, and Francesca Pan. "The Analysis of Extreme Pollution Levels: A Case Study." *Journal of Applied Statistics* 23, no. 2–3 (1996): 333–348.

74 **Dixon had become interested in the topic:** Author interview with Stuart Coles, May 2013.

74 **The work eventually appeared:** Dixon, M. J., and S. G. Coles. "Modelling Association Football Scores and Inefficiencies in the Football Betting Market." *Journal of the Royal Statistical Society: Series C* 46 (1997): 2.

74 **"It was one of those things":** Author interview with Stuart Coles, May 2013.

75 **a single soccer match was so dominated by chance:** Dixon and Coles, "Modelling Association Football Scores."

75 **Researchers have used the Poisson process:** Rakocevic, G., T. Djukic, N. Filipovic, and V. Milutinovic. *Computational Medicine in Data Mining and Modeling* (New York: Springer-Verlag, 2013), 154.

76 **A 2009 study by Christoph Leitner and colleagues:** Leitner, C., A. Zeileis, and K. Hornik. "Forecasting Sports Tournaments by Ratings of (Prob)Abilities: A Comparison for the EURO 2008." *International Journal of Forecasting* 26, no. 3 (2009): 471–481.

77 **Coles would join Smartodds:** Author interview with Stuart Coles, May 2013.

78 **"Those papers are still the main starting points":** Author interview with David Hastie, March 2013.

78 **"It's not an entirely polished piece":** Author interview with Stuart Coles, May 2013.

78 **according to Andreas Heuer and Oliver Rubner:** Heuer, Andreas, and Oliver Rubner. "How Does the Past of a Soccer Match Influence Its Future? Concepts and Statistical Analysis." *PLoS ONE* 7, no. 11 (2012). doi:10.1371/journal.pone.0047678.

79 **"A team would beat another team 28–12":** Details of earlier career come from author interview with Michael Kent, October 2013.

79 **"You need to make your own number":** Author interview with Michael Kent, October 2013.

80 **Because the society's acronym was SABR:** Society for American Baseball Research. "A Guide to Sabermetric Research." http://sabr.org/sabermetrics.

81 **He came to rely on valet parking:** Thomsen, Ian. "The Gang That Beat Las Vegas." *National Sports Daily*, 1990.

82 **There were FBI raids:** Ibid.

82 **the Computer Group placed over $135 million:** Ibid.

83 **"It is a totally different scheme":** Ulam, S. M. *Adventures of a Mathematician* (Oakland: University of California Press, 1991), 311.

83 **It was called the "Avtomat Kalashnikova":** Trex, Ethan. "What Made the AK-47 So Popular?" *Mental Floss*, April 2011. http://mentalfloss.com/article/27455/what-made-ak-47-so-popular.

83 **The gun is still in use today:** Killicoat, Phillip. "Weaponomics: The Global Market for Assault Rifles" (World Bank Policy Research Working Paper 4202, Washington, DC, April 2007).

83 **Complexity means more friction:** Da Silveira, M., L. Gertz, A. Cervieri, A. Rodrigues, et al. "Analysis of the Friction Losses in an Internal Combustion Engine" (SAE Technical Paper 2012–36–0303, 2012). doi:10.4271/2012–36–0303.

84 **US President Woodrow Wilson once described golf:** "A 'Sissy Game' Was the Sport of Presidents." *Life Magazine*, July 1968, 72.

84 **If the ball hits a random object:** Mella, Mirio. "Success = Talent + Luck." Pinnacle Sports, July 15, 2015. http://www.pinnaclesports.com/en/betting-articles/golf/success-talent-luck.

85 **Teams playing in the NHL score:** ESPN. NHL page. http://espn.go.com/nhl/.

85 **where NBA teams will regularly score:** ESPN. NBA page. http://espn.go.com/nba/.

85 **stats such as the "Corsi rating":** Macdonald, Brian. "An Expected Goals Model for Evaluating NHL Teams and Players" (paper presented at MIT Sloan Sports Analytics Conference, Boston, MA, March 2–3, 2012).

85 **the nature of scoring in basketball was changing:** Predictive Sports Betting. MIT Sloan Sports Analytics Conference, Boston, MA,

March 1–2, 2013. Panel discussion with: Chad Millman, Haralabos Voulgaris, and Matthew Holt. Moderator: Jeff Ma. http://www.sloansportsconference.com/?p=9607.

86 **Sifting through reams of data:** Eden, Scott. "Meet the World's Top NBA Gambler." ESPN, February 25, 2013. http://espn.go.com/blog/playbook/dollars/post/_/id/2935/meet-the-worlds-top-nba-gambler.

86 **"We don't realise how easy we have":** Author interview with Stuart Coles, May 2013.

86 **people created automated programs:** Author interview with David Hastie, March 2013.

86 **Some installed countermeasures:** Ward, Mark. "Screen Scraping: How to Profit from Your Rival's Data." BBC News, September 30, 2013. http://www.bbc.com/news/technology-23988890.

86 **"You get a huge database":** Author interview with Michael Kent, October 2013.

87 **That company was Cantor Gaming:** Craig, Susanne. "Taking Risks, Making Odds." *New York Times*, December 24, 2010. http://dealbook.nytimes.com/2010/12/24/taking-risks-making-odds/.

87 **The room feels like a hybrid of a sports bar:** Author experience.

87 **The numbers on Cantor's screens:** Midas background from: Kaplan, Michael. "Wall Street Firm Uses Algorithms to Make Sports Betting Like Stock Trading." *Wired*, November 1, 2010. http://www.wired.com/2010/11/ff_midas/.

87 **Garrood simply went from designing models:** Eden, "Meet the World's Top NBA Gambler."

87 **Cantor wasn't just interested in its predictions:** Craig, "Taking Risks, Making Odds."

88 **Garrood has found that most plays:** Garrood comments from "Betting After the Games Are Underway." ThePostGame, January 11, 2011. http://www.thepostgame.com/blog/spread-sheet/201101/betting-after-games-are-underway.

88 **"We make lines in anticipation":** Predictive Sports Betting, MIT Sloan Sports Analytics Conference, 2013.

89 **"The proprietary nature of prediction models":** McHale, Ian. "Why Spain Will Win . . . Maybe?" *Engineering & Technology*, 5 (June 2010): 25–27.

90 **"There is a lot of paranoia":** Author interview with Rob Esteva, March 2013.

90 **After some suspicious bowling in cricket matches:** Khan, M. Ilyas. "Pakistan's Murky Cricket-Fixing Underworld." BBC News, November 3, 2011. http://www.bbc.com/news/world-asia-15576065.

90 **The scandals have since continued:** Hoult, Nick. "Indian Premier League in Crisis After Three Players Are Charged with Spot Fixing." *Telegraph*, May 16, 2013. http://www.telegraph.co.uk/sport/cricket/twenty20/ipl/10060988/Indian-Premier-League-in-crisis-after-three-players-are-charged-with-spot-fixing.html.

90 **UK police arrested six soccer players:** Hart, Simon. "DJ Campbell Arrested in Connection with Football Fixing." *Telegraph*, December 9, 2013. http://www.telegraph.co.uk/sport/football/10505343/DJ-Campbell-arrested-in-connection-with-football-fixing.html.

90 **the total amount wagered can approach $3 billion:** Wilson, Bill. "World Sport 'Must Tackle Big Business of Match Fixing.'" BBC News, November 25, 2013. http://www.bbc.com/news/business-24984787.

91 **This is the "gray market":** Hawkins, Ed. "Grey Betting Market in Asia Offers Loophole to Be Exploited." *Times* (London), November 30, 2013. http://hawkeyespy.blogspot.com/2013/11/grey-betting-market-in-asia-offers.html.

91 **Haralabos Voulgaris has complained:** Predictive Sports Betting. MIT Sloan Sports Analytics Conference.

91 **Pinnacle claimed it was happy:** Beyer, Andrew. "After Pinnacle, It's All Downhill from Here." *Washington Post*, January 17, 2007. http://www.washingtonpost.com/wp-dyn/content/article/2007/01/16/AR2007011601375.html.

91 **By accepting wagers from sharp bettors:** Noble, Simon. "Inside the Wagering Line." Pinnacle Pulse (blog), Sports Insights, February 22, 2006. https://www.sportsinsights.com/sports-betting-articles/pinnacle-pulse/the-pinnacle-pulse-2222006/.

92 **US Senators stumbled across a Department of Defense proposal:** Taylor, Elanor. "Policy Analysis Market and the Political Yuck Factor." Social Issues Research Centre, April 2004. http://www.sirc.org/articles/policy_analysis.shtml.

92 **One called the idea "grotesque":** Tran, Mark. "Pentagon Scraps Terror Betting Plans." *Guardian*, July 29, 2003. http://www.theguardian.com/world/2003/jul/29/iraq.usa1.

92 **According to Hillary Clinton:** Taylor, "Policy Analysis Market."

93 **Pinnacle has so much faith in the approach:** Wise, Gary. "Head of Sportsbook Q&A Transcript." Pinnacle Sports, August 8, 2013. http://www.pinnaclesports.com/en/betting-articles/social-media/question-answers-with-pinnacle-sports.

93 **Pinnacle dropped horse racing:** "Pinnacle Sports Halts US Horse Racing Service." Casinomeister, December 19, 2008. http://www.casinomeister.com/news/december2008/online_casino_news3/PINNACLE-SPORTS-HALTS-US-HORSE-RACING-SERVICE.php.

93 **Perhaps the best-known betting exchange:** Read, J. J., and J. Goddard. "Information Efficiency in High-Frequency Betting Markets." In *The Oxford Handbook of the Economics of Gambling*, ed. L. V. Williams and D. S. Siegel (New York: Oxford University Press, 2014).

94 **"The boredom was horrendous":** Bowers, Simon. "Odds-on Favourite." *Guardian*, June 6, 2003. http://www.theguardian.com/business/2003/jun/07/9.

94 **the company arranged for a mock funeral procession:** Clarke, Jody. "Andrew Black: Punter Who Revolutionised Gambling." *Moneyweek*, August 21, 2009. http://moneyweek.com/entrepreneurs-my-first-million-andrew-black-betfair-44933/.

94 **if someone wanted to place a bet of £1,000:** Ibid.

96 **"If I find a guy who is good at sports betting":** Klein, Matthew. "Hedge Funds Are Not Necessarily for Suckers." BloombergView, July 12, 2013. http://www.bloombergview.com/articles/2013–07–12/hedge-funds-are-not-necessarily-for-suckers.

96 **According to Tobias Preis:** Preis, Tobias, Dror Y. Kenett, H. Eugene Stanley, Dirk Helbing, and Eshel Ben-Jacob. "Quantifying the Behavior of Stock Correlations Under Market Stress." *Scientific Reports* 2 (2012). doi:10.1038/srep00752.

97 **persuaded Brendan Poots to set up:** Details and quotes from author interview with Brendan Poots, September 2013.

97 **Mark Dixon turned his attention:** Dixon, M. J., and M. E. Robinson. "A Birth Process Model for Association Football Matches." *The Statistician* 47, no. 3 (1998).

98 **you're far more likely to die in a bathtub:** Bailey, Ronald. "How Scared of Terrorism Should You Be?" *Reason Magazine*, September 6, 2011. http://reason.com/archives/2011/09/06/how-scared-of-terrorism-should.

98 **playing roulette again and again:** Spiegelhalter, David. "What's the Best Way to Win Money: Lottery or Roulette?" BBC News, October 14, 2011. http://www.bbc.com/news/uk-15309953.

98 **one 2014 study reckoned the rate:** Titman, A. C., D. A. Costain, P. G. Ridall, and K. Gregory. "Joint Modelling of Goals and Bookings in Association Football." *Journal of the Royal Statistical Society: Series A*, July 15, 2014. doi:10.1111/rssa.12075.

100 **This involves betting on ten soccer leagues:** Author interview with Will Wilde, May 2015.

100 **investment firm Centaur launched the Galileo fund:** Rovell, Darren. "Sports Betting Hedge Fund Becomes Reality." CNBC, April 7, 2010. http://www.cnbc.com/id/36218041.

101 **less than 1 percent of sports bets:** Bell, Kay. "Taxes on Gambling Winnings in Sports." Bankrate, January 2014. http://www.bankrate.com/finance/taxes/taxes-on-gambling-winnings-in-sports-1.aspx.

101 **A new bill, submitted in April 2015:** Takahashi, Maiko. "Japan Lawmakers Group Submits Legislation to Legalize Casinos." Bloomberg Business, April 28, 2015. http://www.bloomberg.com/news/articles/2015–04–28/japan-lawmakers-group-submits-legislation-to-legalize-casinos.

101 **New opportunities will also arise:** Author interview with Will Wilde, May 2015.

102 **Millman got talking to Mike Wohl:** Millman, Chad. "A New System to Bet College Football" (paper presented at MIT Sloan Sports Analytics Conference, Boston, MA, March 1–2, 2013).

102 **"I would want to have play-by-play data":** Author interview with Michael Kent, October 2013.

103 **"We do analysis on the effect":** Author interview with Will Wilde, May 2015.

103 **how a horse's speed changes:** Kaplan, Michael. "The High Tech Trifecta." *Wired*, no. 10.03 (March 2002). http://archive.wired.com/wired/archive/10.03/betting.html.

103 **These "video variables":** Dougherty, Tim. "Horse Sense: Using Applied Mathematics to Game the System." *Contingencies*, May/June 2009. http://www.contingenciesonline.com/contingenciesonline/20090506/?sub_id=qxyLfphSqUiJ#pg22.

103 **Paolo Maldini averaged one tackle:** Kuper, Simon. "How the

Spreadsheet-Wielding Geeks Are Taking over Football." *New Statesman*, June 5, 2013. http://www.newstatesman.com/culture/2013/06/how-spreadsheet-wielding-geeks-are-taking-over-football.

103 **the best cornerbacks in the NFL:** Author interview with Rob Esteva, March 2013.

104 **they won only 52 percent of the games:** Soccer statistics from: Ingle, Sean. "Why the Power of One Is Overhyped in Football." TalkingSport (blog), *Guardian*, March 24, 2013. http://www.theguardian.com/football/blog/2013/mar/24/gareth-bale-one-man-team-overhyped.

104 **"In the sum of the parts":** Author interview with Brendan Poots, September 2013.

105 **"There is a common perception that betting":** Author interview with David Hastie, March 2013.

105 **"We had a guy in New York City":** Author interview with Michael Kent, 2013.

105 **team managers chat with statisticians and modelers:** Eden, "Meet the World's Top NBA Gambler."

105 **The "*Sports Illustrated* jinx":** Wolff, Alexander. "That Old Black Magic." *Sports Illustrated*, January 21, 2002. http://www.si.com/vault/2002/01/21/317048/that-old-black-magic-millions-of-superstitious-readers—and-many-athletes—believe-that-an-appearance-on-sports-illustrateds-cover-is-the-kiss-of-death-but-is-there-really-such-a-thing-as-the-si-jinx.

106 **When a club signs a new player:** McHale, Ian, and Łukasz Szczepański. "A Mixed Effects Model for Identifying Goal Scoring Ability of Footballers." *Journal of the Royal Statistical Society: Series* A 177, no. 2 (2014): 397–417. doi:10.1111/rssa.12015.

106 **Statistician James Albert has attempted:** Albert, James. "Pitching Statistics, Talent and Luck, and the Best Strikeout Seasons of All-Time." *Journal of Quantitative Analysis in Sports* 2, no. 1 (2011).

106 **researchers at Smartodds and the University of Salford:** McHale and Szczepański, "Mixed Effects Model."

107 **Erroneous odds are less common:** Author interview with David Hastie, March 2013.

107 **"I would start with the minor sports":** Predictive Sports Betting, MIT Sloan Sports Analytics Conference.

Chapter 5

109 **"What hath God wrought!":** History of the telegram comes from: "The Birth of Electrical Communications—1837." University of Salford. http://www.cntr.salford.ac.uk/comms/ebirth.php.

110 **traders used telegrams to tell each other:** Poitras, Geoffrey. "Arbitrage: Historical Perspectives." *Encyclopedia of Quantitative Finance*, 2010. doi:10.1002/9780470061602.eqf01010.

110 **traders refer to the GBP/USD exchange rate:** Author experience.

110 **Some would even trek further afield:** Poitras, "Arbitrage: Historical Perspectives."

111 **researchers at Athens University looked at bookmakers' odds:** Vlastakis, Nikolaos, George Dotsis, and Raphael N. Markellos. "How Efficient Is the European Football Betting Market? Evidence from Arbitrage and Trading Strategies." *Journal of Forecasting* 28, no. 5 (2009): 426–444.

111 **a group at the University of Zurich searched:** Franck, Egon, Erwin Verbeek, and Stephan Nüesch. "Inter-market Arbitrage in Sports Betting" (NCER Working Paper Series no. 48, National Centre for Econometric Research, Brisbane, Queensland, Australia, October 2009). http://www.ncer.edu.au/papers/documents/WPNo48.pdf.

111 **Economist Milton Friedman pointed out:** Beinhocker, Eric. *The Origin of Wealth: Evolution, Complexity, and the Radical Remaking of Economics* (Cambridge, MA: Harvard Business Press, 2006), 396.

112 **a group of researchers at the University of Lancaster:** Buraimo, Babatunde, David Peel, and Rob Simmons. "Gone in 60 Seconds: The Absorption of News in a High-Frequency Betting Market" (working paper, from the Selected Works of Dr. Babatunde Buraimo, March 2008). http://works.bepress.com/babatunde_buraimo/17.

113 **Of the 4.4 million bets placed:** "Backing a Winner." *Computing Magazine*, January 25, 2007. http://www.computing.co.uk/ctg/analysis/1854505/backing-winnerw.

113 **"These algorithms mop up any mispricing":** Author interview with David Hastie, March 2013.

113 **It currently takes 65 milliseconds:** Williams, Christopher. "The $300m Cable That Will Save Traders Milliseconds." *Telegraph*,

September 11, 2011. http://www.telegraph.co.uk/technology/news/8753784/The-300m-cable-that-will-save-traders-milliseconds.html.

113 **one blink of the human eye:** Tucker, Andrew. "In the Blink of an Eye." Optalert, August 5, 2014. http://www.optalert.com/news/in-the-blink-of-an-eye.

114 **Traders call the problem "slippage":** Liberty, Jez. "Measuring and Avoiding Slippage." *Futures Magazine*, August 1, 2011. http://www.futuresmag.com/2011/07/31/measuring-and-avoiding-slippage.

115 **The resulting trade is known as an "iceberg order":** Almgren, Robert, and Bill Harts. "Smart Order Routing" (StreamBase White Paper, 2008). http://www.streambase.com/wp-content/uploads/downloads/StreamBase_White_Paper_Smart_Order_Routing_low.pdf.

115 **One example is a "sniffing algorithm":** Ablan, Jennifer. "Snipers, Sniffers, Guerillas: The Algo-Trading War." Reuters, May 31, 2007. http://www.reuters.com/article/2007/05/31/businesspro-usa-algorithm-strategies-dc-idUSN3040797620070531.

116 **As the hooves pounded the ground:** Details from: Rushton, Katherine. "Betfair Loses £40m on Leopardstown After 'Technical Glitch.'" *Telegraph*, December 29, 2011. http://www.telegraph.co.uk/finance/newsbysector/retailandconsumer/8983469/Betfair-loses-40m-on-Leopardstown-after-technical-glitch.html.

116 **Soon after the race finished:** Betfair forum thread: "Hope you all took advantage of betfairs xmas bonus." Geeks Toy Horseracing forum, December 28, 2011. http://www.geekstoy.com/forum/showthread.php?7065-Hope-you-all-took-advantage-of-betfairs-xmas-bonus.

117 **"Due to a technical glitch":** Webb, Peter. "£1k Account Caused £600m Betfair Error." Bet Angel Blog, December 2011. http://www.betangel.com/blog_wp/2011/12/30/1k-account-caused-600m-betfair-error/.

117 **"You cannot win—or lose":** Wood, Greg. "Betfair May Lose Out by Not Explaining How £600m Lay Bet Was Accepted." Talking Sport (blog), *Guardian*, December 30, 2011. http://www.theguardian.com/sport/blog/2011/dec/30/betfair-600m-lay-bet.

117 **The summer of 2012 was a busy time:** Details of the events come from: SEC report. "In the Matter of Knight Capital Americas LLC." File No. 3–15570. October 2013.

119 **In 2007, a trader named Svend Egil Larsen:** Details of the Larsen case come from: Stothard, Michael. "Day Traders Expose Algorithm's

Flaws." *Globe and Mail*, May 16, 2012. http://www.theglobeandmail.com/globe-investor/day-traders-expose-algorithms-flaws/article4179395/; and Stothard, Michael. "Norwegian Day Traders Cleared of Wrongdoing." *Financial Times*, May 2, 2012. http://www.ft.com/cms/s/0/e2f6d1cc-9447–11e1-bb47–00144feab49a.html#axzz3hDw6Bgnj.

120 **Farmer has pointed out:** Farmer, J. Doyne, and Duncan Foley. "The Economy Needs Agent-Based Modelling." *Nature* 460 (2009): 685–686. doi:10.1038/460685a.

120 **At lunchtime on April 23, 2013:** Foster, Peter. "'Bogus' AP Tweet About Explosion at the White House Wipes Billions off US Markets." *Telegraph*, April 23, 2013. http://www.telegraph.co.uk/finance/markets/10013768/Bogus-AP-tweet-about-explosion-at-the-White-House-wipes-billions-off-US-markets.html.

121 **One of the biggest market shocks:** Details of the flash crash come from: US Commodity Futures Trading Commission and US Securities and Exchange Commission. *Findings Regarding the Market Events of May 6, 2010.* September 30, 2010. https://www.sec.gov/news/studies/2010/marketevents-report.pdf.

122 **Algorithms sift through the reports:** Sonnad, Nikhil. "The AP's Newest Business Reporter Is an Algorithm." *Quartz*, June 30, 2014. http://qz.com/228218/the-aps-newest-business-reporter-is-an-algorithm/.

122 **To understand the problem:** Keynes, John M. *The General Theory of Employment, Interest, and Money* (London: Palgrave Macmillan, 1936).

123 **"As soon as you limit what you can do":** Quotes come from author interview with J. Doyne Farmer, October 2013.

124 **Some traders have reported:** Farrell, Maureen. "Mini Flash Crashes: A Dozen a Day." CNN Money. March 20, 2013. http://money.cnn.com/2013/03/20/investing/mini-flash-crash/.

124 **they found thousands of "ultrafast extreme events":** Johnson, Neil, Guannan Zhao, Eric Hunsader, Hong Qi, Nicholas Johnson, Jing Meng and Brian Tivnan. "Abrupt Rise of New Machine Ecology Beyond Human Response Time." *Scientific Reports* 3 (2013). doi:10.1038/srep02627.

124 **"Humans are unable to participate in real time":** Quote from: "Robots Take Over Economy: Sudden Rise of Global Ecology of Interacting Robots Trade at Speeds Too Fast for Humans" (press release, University of Miami, September 11, 2013).

125 **The campus is a maze of neo-Gothic halls:** Author experience.

125 **"The trees on the right were passing me":** Halmos, Paul. "The Legend of John von Neumann." *American Mathematical Monthly* 8 (1973): 382–394.

125 **To examine how different factors influenced ecological systems:** Details of model from: May, R. M. "Simple Mathematical Models with Very Complicated Dynamics." *Nature* 261 (1976): 459–467.

125 **This was first proposed in 1838:** Bacaër, Nicolas. "Verhulst and the Logistic Equation (1838)." *A Short History of Mathematical Population Dynamics* (2011): 35–39.

128 **May found that the larger the ecosystem:** May, Robert M. "Will a Large, Complex Ecosystem Be Stable?" *Nature* 238 (1972): 413–414. doi:10.1038/238413a0.

129 **According to ecologist Andrew Dobson:** Dobson, Andrew. "Multi-Host, Multi-Parasite Dynamics" (Infectious Disease Dynamics workshop, Isaac Newton Institute, Cambridge, UK,. August 19–23, 2013).

129 **Yet, according to Stefano Allesina and Si Tang:** Allesina, Stefano, and Si Tang. "Stability Criteria for Complex Ecosystems." *Nature* 483 (2012): 205–208. doi:10.1038/nature10832.

131 **Doyne Farmer has pointed out:** Farmer, J. Doyne. "Market Force, Ecology and Evolution." *Industrial and Corporate Change* 11, no. 5 (2002): 895–953.

132 **One of the most popular types of financial wager:** Investment Trends. *2013 UK Leveraged Trading Report.* December 23, 2013. http://www.iggroup.com/content/files/leveraged_trading_report_nov13.pdf.

133 **If you make a profitable stock trade:** HM Revenue and Customs. "General Betting Duty." 2010. https://www.gov.uk/general-betting-duty.

133 **In Australia, profits from spread betting:** Armitstead, Louise. "Treasury to Look at Spread Betting Tax Exemption After Lords Raise Concerns." *Telegraph*, November 27, 2013. http://www.telegraph.co.uk/finance/newsbysector/banksandfinance/10479460/Treasury-to-look-at-spread-betting-tax-exemption-after-Lords-raise-concerns.html.

133 **In 2006, the US Federal Reserve:** Details from: "New Directions for Understanding Systemic Risk" (report on a conference cosponsored by the Federal Reserve Bank of New York and the National Academy of Sciences, New York, NY, May 2006).

CHAPTER 6

135 **In summer 2010, poker websites launched:** Dance, Gabriel. "Poker Bots Invade Online Gambling." *New York Times*, March 13, 2011. http://www.nytimes.com/2011/03/14/science/14poker.html.

135 **Swedish police started investigating poker bots:** Wood, Jocelyn. "Police Investigating Coordinated Poker Bot Operation in Sweden." Pokerfuse, February 22, 2013. http://pokerfuse.com/news/poker-room-news/police-investigating-million-dollar-poker-bot-operation-sweden-21–02/.

135 **It turned out that these bots:** Jones, Nick. "Over $500,000 Repaid to Victims of Bot Ring on Svenska Spel." Pokerfuse, June 20, 2013. http://pokerfuse.com/news/poker-room-news/over-500000-repaid-to-victims-of-bot-ring-on-svenska-spel/.

135 **Until these sophisticated computer players:** Ruddock, Steve. "Alleged Poker Bot Ring Busted on Swedish Poker Site." Poker News Boy, February 24, 2013. http://pokernewsboy.com/online-poker-news/alleged-poker-bot-ring-busted-on-swedish-poker-site/13633.

136 **this was an industry that had spent over $300 million:** Surgeon General. *Preventing Tobacco Use Among Youth and Young Adults: A Report of the Surgeon General, 2012* (Washington, DC: National Center for Chronic Disease Prevention and Health Promotion Office on Smoking and Health, 2012), Table 5.3.

136 **The vote was scheduled:** McGrew, Jane. "History of Tobacco Regulation." In *Marihuana: A Signal of Misunderstanding* (report of the National Commission on Marihuana and Drug Abuse, 1972). http://www.druglibrary.org/schaffer/library/studies/nc/nc2b.htm.

136 **Far from hurting tobacco companies' profits:** McAdams, David. *Game-Changer: Game Theory and the Art of Transforming Strategic Situations* (New York: W. W. Norton, 2014), 61.

137 **Yet tobacco revenues held steady:** Hamilton, James. "The Demand for Cigarettes: Advertising, the Health Scare, and the Cigarette Advertising Ban." *Review of Economics and Statistics* 54, no. 4 (1972).

137 **"Mr. Nash is nineteen years old":** The letter was posted online by Princeton University after John Nash's death in 2015. It went viral.

138 **Despite his prodigious academic record:** Halmos, Paul. "The Legend of John von Neumann." *American Mathematical Monthly* 8 (1973): 382–394.

138 **"Real life consists of bluffing":** Harford, Tim. "A Beautiful Theory." *Forbes*, December 14, 2006. http://www.forbes.com/2006/12/10/business-game-theory-tech-cx_th_games06_1212harford.html. Original quote made in BBC show "Ascent of Man," broadcast in 1973.

138 **Von Neumann started by looking at poker:** Ferguson, Chris, and Thomas S. Ferguson. "On the Borel and von Neumann Poker Models." *Game Theory and Applications* 9 (2003): 17–32.

139 **in a book titled *Theory of Games and Economic Behavior*:** Von Neumann, John, and Oskar Morgenstern. *Theory of Games and Economic Behavior* (Princeton, NJ: Princeton University Press, 1944).

139 **Despite his fondness for Berlin's nightlife:** Dyson, Freeman. "A Walk Through Johnny von Neumann's Garden." *Notices of the AMS* 60, no. 2 (2010): 154–161.

140 **So, it was only natural:** Las Vegas: An Unconventional History. "Benny Binion (1904–1989)." PBS.org, 2005. http://www.pbs.org/wgbh/amex/lasvegas/peopleevents/p_binion.html.

140 **Early in the 1982 competition:** Monroe, Billy. "Where Are They Now—Jack Straus." Poker Works, April 11, 2008. http://pokerworks.com/poker-news/2008/04/11/where-are-they-now-jack-straus.html.

140 **the thirty-first World Series reached its finale:** Details come from video of final at: http://www.tjcloutierpoker.net/2000-world-series-of-poker-final-table-chris-ferguson-vs-tj-cloutier/. TJ Cloutier Poker. "2000 World Series of Poker Final Table—Chris Ferguson vs TJ Cloutier." October 12, 2010.

141 **"You didn't think it would be that tough":** Paulle, Mike. "If You Build It They Will Come." *ConJelCo* 31, no. 25 (May 14–18, 2000). http://www.conjelco.com/wsop2000/event27.html.

141 **no poker player had won more than $1 million:** Wilkinson, Alec. "What Would Jesus Bet?" *The New Yorker*, March 30, 2009. http://www.newyorker.com/magazine/2009/03/30/what-would-jesus-bet.

141 **consultant for the California State Lottery:** Johnson, Linda. "Chris Ferguson, 2000 World Champion." *CardPlayer Magazine* 16, no. 18 (2003).

142 **Combined with improvements in computing power:** Details from: Wilkinson, "What Would Jesus Bet?"

142 **Building on von Neumann's ideas:** Ferguson, C., and T. Ferguson. "The Endgame in Poker." In *Optimal Play: Mathematical Studies of Games and Gambling*, ed. Stewart N. Ethier and William R. Eadington

(Reno, NV: Institute for the Study of Gambling and Commercial Gaming, 2007).

143 **"You always want to make your opponents' decisions":** Ferguson, Chris. "Sizing Up Your Opening Bet." Hendon Mob, October 7, 2007. http://www.thehendonmob.com/poker_tips/sizing_up_your_opening_bet_by_chris_ferguson.

143 **As well as winning more money:** Harford, "Beautiful Theory."

143 **"How do I win the most?":** Wilkinson, "What Would Jesus Bet?"

144 **He once taught himself:** Johnson, "Chris Ferguson."

144 **Starting with nothing:** Details of challenge from: Ferguson, Chris. "Chris Ferguson's Bankroll Challenge." PokerPlayer, March 2009. http://www.pokerplayer365.com/poker-players/player-interviews-poker-players/read-about-chris-fergusons-bankroll-challenge-and-you-could-turn-0-into-10000/.

144 **"I remember winning my first $2":** Ferguson. "Chris Ferguson's Bankroll Challenge."

146 **When Ignacio Palacios-Heurta:** Palacios-Heurta, Ignacio. "Professionals Play Minimax." *Review of Economic Studies* 70 (2003): 395–415.

147 **Von Neumann completed his solution:** Details of the dispute were given in: Kjedldsen, T. H. "John von Neumann's Conception of the Minimax Theorem: A Journey Through Different Mathematical Contexts." *Archive for History of Exact Science* 56 (2001).

149 **While earning his master's degree in 2003:** Follek, Robert. "SoarBot: A Rule-Based System for Playing Poker" (MSc diss., School of Computer Science and Information Systems, Pace University, 2003).

150 **Led by David Hilbert:** O'Connor, J. J., and E. F. Robertson. "Biography of John von Neumann." *JOC/EFR*, October 2003. http://www-history.mcs.st-and.ac.uk/Biographies/Von_Neumann.html.

150 **some inconsistencies in the US Constitution:** "Kurt Gödel." Institute for Advanced Study Online, 2013. https://www.ias.edu/people/godel.

151 **poker bots grew in popularity:** Kushner, David. "On the Internet, Nobody Knows You're a Bot." *Wired* 13.09 (September 2005). http://archive.wired.com/wired/archive/13.09/pokerbots.html?tw=wn_tophead_7.

151 **Just as stripped-down versions of poker:** Details of strategies given

in: Rubin, Jonathan, and Ian Watson. "Computer Poker: A Review." *Artificial Intelligence* 175 (2011): 958–987.

152 **technique known as "regret minimization":** Ibid.

153 **In 2000, researchers at the University of Iowa reported:** Bechara, A., Hanna Damasio, and Antonio R. Damasio. "Emotion, Decision Making and the Orbitofrontal Cortex." *Cerebral Cortex* 10, no. 3 (2000): 295–307. doi:10.1093/cercor/10.3.295.

153 **This contrasts with much economic theory:** Cohen, Michael D. "Learning with Regret." *Science* 319, no. 5866 (2008): 1052–1053.

154 **at the University of Alberta in Canada:** Schaeffer, Jonathan. "Marion Tinsley: Human Perfection at Checkers?" http://www.wylliedraughts.com/Tinsley.htm.

155 **The name was a pun:** Propp, James. "Chinook." *ACJ Extra*, 1999. http://faculty.uml.edu/jpropp/chinook.html.

155 **That's 10 followed by twenty zeros:** Estimate given in: Mackie, Glen. "To See the Universe in a Grain of Taranaki Sand." *North and South Magazine*, May 1999. http://astronomy.swin.edu.au/~gmackie/billions.html.

155 **Chinook "pruned" this decision tree:** Details of competition in Schaeffer, Jonathan, Robert Lake, Paul Lu, and Martin Bryant. "Chinook: The World Man-Machine Checkers Champion." *AI Magazine* 17, no. 1 (1996). doi:http://dx.doi.org/10.1609/aimag.v17i1.1208.

156 **he coined the infinite monkey theorem:** Borel, E. M. "La mécanique statique et l'irréversibilité." *Journal of Theoretical and Applied Physics*, 1913.

158 **"Checkers is solved":** Schaeffer, Jonathan, Neil Burch, Yngvi Björnsson, Akihiro Kishimoto, Martin Müller, Robert Lake, Paul Lu, and Steve Sutphen. "Checkers Is Solved." *Science* 317, no. 5844 (2007): 1518–1522. doi:10.1126/science.1144079.

158 **John Nash showed in 1949:** Demaine, Erik D., and Robert A. Hearn. "Playing Games with Algorithms: Algorithmic Combinatorial Game Theory." *Mathematical Foundations of Computer Science* (2001): 18–32. http://erikdemaine.org/papers/AlgGameTheory_GONC3/paper.pdf.

159 **Twenty-six moves later:** Schaeffer, Jonathan, and Robert Lake. "Solving the Game of Checkers." *Games of No Chance* 29 (1996): 119–133. http://library.msri.org/books/Book29/files/schaeffer.pdf.

160 **"might have died in 1990":** Schaeffer et al., "Chinook."

160 **Doyne Farmer has started to question:** Galla, Tobias, and J. Doyne Farmer. "Complex Dynamics in Learning Complicated Games." *PNAS* 110, no. 4 (2013): 1232–1236. doi:10.1073/pnas.1109672110.

162 **"Large changes tend to be followed":** Mandelbrot, Benoit. "The Variation of Certain Speculative Prices." *Journal of Business* 36, no. 4 (1963): 394–419. http://www.jstor.org/stable/2350970.

163 **"You have a very strong program":** Billings, D., N. Burch, A. Davidson, R. Holte, J. Schaeffer, T. Schauenberg, and D. Szafro. "Approximating Game-Theoretic Optimal Strategies for Full-Scale Poker." *IJCAI* (2003): 661–668. http://ijcai.org/Past%20Proceedings/IJCAI-2003/PDF/097.pdf.

CHAPTER 7

165 **Thanks to their ability to dissect:** Background on Watson comes from: Rashid, Fahmida. "IBM's Watson Ties for Lead on *Jeopardy* but Makes Some Doozies." EWeek, February 14, 2011. http://www.eweek.com/c/a/IT-Infrastructure/IBMs-Watson-Ties-for-Lead-on-Jeopardy-but-Makes-Some-Doozies-237890; and Best, Jo. "IBM Watson: How the *Jeopardy*-Winning Supercomputer Was Born, and What It Wants to Do Next." TechRepublic. http://www.techrepublic.com/article/ibm-watson-the-inside-story-of-how-the-jeopardy-winning-supercomputer-was-born-and-what-it-wants-to-do-next/.

166 **IBM collected some of the results:** Basulto, Dominic. "How IBM Watson Helped Me to Create a Tastier Burrito Than Chipotle." *Washington Post*, April 15, 2015. http://www.washingtonpost.com/blogs/innovations/wp/2015/04/15/how-ibm-watson-helped-me-to-create-a-tastier-burrito-than-chipotle/.

167 **"Let's try poker":** Wise, Gary. "Representing Mankind." ESPN Poker Club, August 6, 2007. http://sports.espn.go.com/espn/poker/columns/story?columnist=wise_gary&id=2959684.

167 **Finally, there is Eric Jackson:** Details and quotes from author interviews with Michael Johanson and Neil Burch, April 2014, and Tuomas Sandholm, December 2013. Additional specifics from competition online results (http://www.computerpokercompetition.org).

168 **"Poker is a perfect microcosm":** Author interview with Jonathan Schaeffer, July 2013.

169 **"a bath of refreshing foolishness":** Ulam, S. M. *Adventures of a Mathematician* (Oakland: University of California Press, 1991).

169 **young British mathematician by the name of Alan Turing:** Hodges, Andrew. *Alan Turing: The Enigma* (Princeton, NJ: Princeton University Press, 1983).

169 **"I rather liked it at first":** Turing background given in: Copeland, B. J. *The Essential Turing* (Oxford: Oxford University Press, 2004).

170 **manuscript entitled "The Game of Poker":** The game of poker. File AMT/C/18. The Papers of Alan Mathison Turing. The UK National Archives.

170 **He also wondered how games:** Details of the imitation game given in: Turing, A. M. "Computing Machinery and Intelligence." *Mind* 59 (1950): 433–460.

171 **When it played chess against Garry Kasparov:** Kasparov, Garry. "The Chess Master and the Computer." *New York Review of Books*, February 11, 2010. http://www.nybooks.com/articles/archives/2010/feb/11/the-chess-master-and-the-computer/.

172 **In 2013, journalist Michael Kaplan:** Details of Vegas bot given in: Kaplan, Michael. "The Steely, Headless King of Texas Hold 'Em." *New York Times Magazine*, September 5, 2013. http://www.nytimes.com/2013/09/08/magazine/poker-computer.html.

173 **It would have to read its opponent:** Comparison of poker and backgammon in: Dahl, Fredrik. "A Reinforcement Learning Algorithm Applied to Simplified Two-Player Texas Hold'em Poker." *EMCL '01 Proceedings of the 12th European Conference on Machine Learning* (2001): 85–96. doi:10.1007/3–540–44795–4_8.

174 **Neural networks are not a new idea:** McCulloch, Warren S., and Walter H. Pitts. "A Logical Calculus of the Ideas Immanent in Nervous Activity." *Bulletin of Mathematical Biophysics* 5 (1943): 115–133. http://www.cse.chalmers.se/~coquand/AUTOMATA/mcp.pdf.

174 **Facebook announced an AI team:** Details of AI team and DeepFace in: Simonite, Tom. "Facebook Launches Advanced AI Effort to Find Meaning in Your Posts." *MIT Technology Review*, September 20, 2013. http://www.technologyreview.com/news/519411/facebook-launches-advanced-ai-effort-to-find-meaning-in-your-posts/; and Simonite, Tom. "Facebook Creates Software That Matches

Faces Almost as Well as You Do." *MIT Technology Review*, March 17, 2014. http://www.technologyreview.com/news/525586/facebook-creates-software-that-matches-faces-almost-as-well-as-you-do/.

174 **Facebook users were uploading over 350 million:** Smith, Cooper. "Facebook Users Are Uploading 350 Million New Photos Each Day." *Business Insider*, September 18, 2013. http://www.businessinsider.com/facebook-350-million-photos-each-day-2013–9.

176 **Rather than grab a vulnerable pawn:** Description of move in: Chelminski, Rudy. "This Time It's Personal." *Wired* 9.10 (October 2001). http://archive.wired.com/wired/archive/9.10/chess.html.

176 **Deep Blue's game-changing show:** Fact that the move was random from: Silver, Nate. *The Signal and the Noise: Why So Many Predictions Fail—but Some Don't* (London: Penguin, 2012).

176 **Some are easier to scare off than others:** Bateman, Marcus. "What Does 'Floating' Mean?" Betfair Online, July 6, 2010. https://betting.betfair.com/poker/poker-strategy/what-does-floating-mean-060710.html.

177 **"Most of our group aren't poker players":** Author interview with Michael Johanson and Neil Burch, April 2014.

178 **In 2010, an online version of rock-paper-scissors:** Dance, Gabriel, and Tom Jackson. "Rock-Paper-Scissors: You vs. the Computer." *New York Times*. http://www.nytimes.com/interactive/science/rock-paper-scissors.html.

178 **In 2014, Zhijian Wang and colleagues:** Wang, Zhijian, Bin Xu, and Hai-Jun Zhou. "Social Cycling and Conditional Responses in the Rock-Paper-Scissors Game." *Scientific Reports* 4, no. 5830 (2014). doi:10.1038/srep05830.

179 **cognitive psychologist George Miller noted:** Miller, George A. "The Magical Number Seven, Plus or Minus Two: Some Limits on Our Capacity for Processing Information." *Psychological Review* 63 (1956): 81–97.

179 **Dutch psychologist Willem Wagenaar observed:** Bar-Hillel, Maya, and Willem A. Wagenaar. "The Perception of Randomness." *Advances in Applied Mathematics* 12, no. 4 (1991): 428–454. doi:10.1016/0196–8858(91)90029-I.

179 **referred to as the "magical number seven":** Jacobson, Roni. "Seven Isn't the Magic Number for Short-Term Memory." *New York Times*, September 9, 2013.

180 **the best competitors can memorize:** Lai, Angel. "World Records." http://www.world-memory-statistics.com/disciplines.php.

180 **memorizing cards also helps in blackjack:** Details about memory techniques in: Robb, Stephen. "How a Memory Champ's Brain Works." BBC News, April 7, 2009. http://news.bbc.co.uk/2/hi/uk_news/magazine/7982327.stm.

180 **"often mused about the nature of memory":** Metropolis, Nick. "The Beginning of the Monte Carlo Method." Special issue, *Los Alamos Science* (1987): 125–130. http://jackman.stanford.edu/mcmc/metropolis1.pdf.

181 **The database came from Shawn Bayern:** "Rock-Paper-Scissors: Humans Versus AI." http://www.essentially.net/rsp.

182 **"A coalition absorbs at least two players":** Von Neumann, J., and Oskar Morgenstern. *Theory of Games and Economic Behavior* (Princeton, NJ: Princeton University Press, 1944).

182 **Parisa Mazrooei and colleagues at the University of Alberta:** Mazrooei, Parisa, Christopher Archibald, and Michael Bowling. "Automating Collusion Detection in Sequential Games." *Association for the Advancement of Artificial Intelligence* (2013). https://webdocs.cs.ualberta.ca/~bowling/papers/13aaai-collusion.pdf.

182 **There are reports of unscrupulous players:** Goldberg, Adrian. "Can the World of Online Poker Chase Out the Cheats?" BBC News, September 12, 2010. http://www.bbc.com/news/uk-11250835.

182 **"In any form of poker":** Dahl, F. "A Reinforcement Learning Algorithm Applied to Simplified Two-Player Texas Hold'em Poker." In *European Conference on Machine Learning 2001, Lecture Notes in Artificial Intelligence 2167*, ed. L. De Raedt and P. Flach (Berlin: Springer-Verlag, 2001).

184 **tweak your tactics as you learn:** Author interview with Tuomas Sandholm, December 2013. Additional details in: Sandholm, T. "Perspectives on Multiagent Learning." *Artificial Intelligence* 171 (2007): 382–391.

184 **Sandholm has been developing "hybrid" bots:** Ganzfried, Sam, and Tuomas Sandholm. "Game Theory-Based Opponent Modeling in Large Imperfect-Information Games." *Proceedings of the 10th International Conference on Autonomous Agents and Multiagent Systems* 2 (2011): 533–540.

185 **professional players Phil Laak and Ali Eslami:** Details of event in: Wise, "Representing Mankind"; and Harris, Martin. "Laak-Eslami

Team Defeats Polaris in Man-Machine Poker Championship." PokerNews, July 25, 2007. http://www.pokernews.com/news/2007/07/laak-eslami-team-defeats-polaris-man-machine-poker-champions.htm.

186 **there was a second man-machine competition:** Details of event in: Harris, Martin. "Polaris 2.0 Defeats Stoxpoker Team in Man-Machine Poker Championship." PokerNews, July 10, 2008. http://www.pokernews.com/news/2008/07/man-machine-II-poker-championship-polaris-defeats-stoxpoker-.htm; and Johnson, R. Colin. "AI Beats Human Poker Champions." EETimes, July 7, 2008. http://www.eetimes.com/document.asp?doc_id=1168863.

187 **Using the regret minimization approach:** Author interview with Michael Johanson, April 2014.

188 **With a nod to the group's checkers research:** Bowling, Michael, Neil Burch, Michael Johanson, and Oskari Tammelin. "Heads-Up Limit Hold'em Poker Is Solved." *Science* 347, no. 6218 (2015): 145–149. doi:10.1126/science.1259433.

189 **"It would attack the mystique":** Author interview with Michael Johanson, April 2014.

190 **Watson found the short clues the most difficult:** Sutton, John D. "Behind-the-Scenes with IBM's 'Jeopardy!' Computer, Watson." CNN, February 7, 2011. http://www.cnn.com/2011/TECH/innovation/02/07/watson.ibm.jeopardy/.

190 **people might be especially good at sizing up others:** Wright, G. R., C. J. Berry, and G. Bird. "'You Can't Kid a Kidder': Association Between Production and Detection of Deception in an Interactive Deception Task." *Frontiers in Human Neuroscience* 6 (2012): 87. doi:10.3389/fnhum.2012.00087.

191 **In a 2006 survey spanning fifty-eight countries:** Global Deception Research Team. "A World of Lies." *Journal of Cross-Cultural Psychology* 37, no. 1 (2006): 60–74. doi:10.1177/0022022105282295.

191 **There's no evidence that liars avert their gaze:** DePaulo, B. M., J. J. Lindsay, B. E. Malone, L. Muhlenbruck, K. Charlton, and H. Cooper. "Cues to Deception." *Psychological Bulletin* 129, no. 1 (2003): 74–118.

191 **In a 2010 study:** Schlicht, E. J., S. Shimojo, C. F. Camerer, P. Battaglia, and K. Nakayama. "Human Wagering Behavior Depends on Opponents' Faces." *PLoS ONE* 5, no. 7 (2010): e11663.

192 **When Matt Mazur decided to build a poker bot:** Author interview with Matt Mazur, August 2014. Additional details from his blog posts (http://www.mattmazur.com).

CHAPTER 8

197 **Hundreds of cameras cling:** Author experience.

197 **casinos' definition of such cheating:** History of surveillance in: Hicks, Jesse. "Not in My House: How Vegas Casinos Wage a War on Cheating." *The Verge*, January 14, 2014. http://www.theverge.com/2013/1/14/3857842/las-vegas-casino-security-versus-cheating-technology.

198 **Unlawful Internet Gambling Enforcement Act:** Unlawful Internet Gambling Enforcement Act of 2006, 31 U.S.C. 5361–5366, §5362.

198 **That included Lawrence DiCristina:** Details of DiCristina case from: Weinstein, Jack. *Memorandum, Order & Judgment, United States of America against Lawrence DiCristina*. 11-CR-414. August 2012. http://jurist.org/paperchase/103482098-U-S-vs-DiCristina-Opinion-08–21–2012.pdf.

200 **airport operator William McBoyle helped arrange:** *McBoyle v. U.S.* 1930 10CIR 118, 43 F.2d 273.

201 **The conviction was reversed:** Paraphrased from original comment in *McBoyle v. U.S.* 1930: "When a rule of conduct is laid down in words that evoke in the common mind only the picture of vehicles moving on land, the statute should not be extended to aircraft simply because it may seem to us that a similar policy applies, or upon the speculation that, if the legislature had thought of it, very likely broader words would have been used."

201 **Rather, the state law meant:** Brennan, John. "U.S. Supreme Court Declines to Take DiCristina Poker Case; Reminder of Challenge Faced by NJ Sports Betting Advocates." NorthJersey.com, February 24, 2014. http://blog.northjersey.com/meadowlandsmatters/7891/u-s-supreme-court-declines-to-take-dicristina-poker-case-reminder-of-challenge-faced-by-nj-sports-betting-advocates/.

202 **"There may be such a thing as habitual luck":** Ulam, S. M. *Adventures of a Mathematician* (Oakland: University of California Press, 1991).

202 **"Chance favours the prepared mind":** Quoted in: Weiss, R. A. "HIV and the Naked Ape." In *Serendipity: Fortune and the Prepared Mind*,

ed. M. De Rond and I. Morley (Cambridge: Cambridge University Press, 2010). Originally said during a lecture at University of Lille, 1854.

203 **Matthew Salganik and colleagues at Columbia University:** Salganik, M. J., P. S. Dodds, and D. J. Watts. "Experimental Study of Inequality and Unpredictability in an Artificial Cultural Market." *Science* 311 (2006): 854–856.

203 **"Fame has much less to do . . .":** Dodds, Peter Sheridan. "Homo Narrativus and the Trouble with Fame." *Nautilus,* September 5, 2013. http://nautil.us/issue/5/fame/homo-narrativus-and-the-trouble-with-fame.

204 **"Consider a set of funds with no skill":** Roulston, Mark, and David Hand. "Blinded by Optimism" (working paper, Winton Capital Management, December 2013). https://www.wintoncapital.com/assets/Documents/BlindedbyOptimism.pdf?1398870164.

205 **hockey analyst Brian King suggested a way:** Charron, Cam. "Analytics Mailbag: Save Percentages, PDO, and Repeatability." TheLeafsNation.com. May 27, 2014. http://theleafsnation.com/2014/5/27/analytics-mailbag-save-percentages-pdo-and-repeatability.

205 **The statistic, later dubbed PDO:** Details on PDO and NHL statistics given in: Weissbock, Joshua, Herna Viktor, and Diana Inkpen. "Use of Performance Metrics to Forecast Success in the National Hockey League" (paper presented at the European Conference on Machine Learning and Principles and Practice of Knowledge Discovery in Databases, Prague, September 23–27, 2013).

205 **England had the lowest PDO:** Burn-Murdoch, John. "Were England the Uunluckiest Team in the World Cup Group Stages?" FT Data Blog. 29 June 2014. http://blogs.ft.com/ftdata/2014/06/29/were-england-the-unluckiest-team-in-the-world-cup-group-stages/.

206 **Cambridge college spent on wine:** "In Vino Veritas, Redux." *The Economist,* February 5, 2014. http://www.economist.com/blogs/freeexchange/2014/02/correlation-and-causation-0.

207 **topped the wine list with a spend of £338,559:** Simons, John. "Wages Not Wine: Booze Hound Colleges Spend £3 million on Wine." *Tab* (Cambridge, England), January 22, 2014. http://thetab.com/uk/cambridge/2014/01/22/booze-hound-colleges-spend-3-million-on-wine-32441.

207 **Countries that consume lots of chocolate:** Messerli, F. H. "Chocolate Consumption, Cognitive Function, and Nobel Laureates." *New England Journal of Medicine* 367 (2012): 1562–1564. doi:10.1056/NEJMon1211064.

207 **When ice cream sales rise in New York City:** Peters, Justin. "When Ice Cream Sales Rise, So Do Homicides. Coincidence, or Will Your Next Cone Murder You?" Crime (blog), *Slate*, July 9, 2013. http://www.slate.com/blogs/crime/2013/07/09/warm_weather_homicide_rates_when_ice_cream_sales_rise_homicides_rise_coincidence.html.

209 **When Manchester City won the league:** Lewis, Tim. "How Computer Analysts Took Over at Britain's Top Football Clubs." *The Observer*, March 9, 2014. http://www.theguardian.com/football/2014/mar/09/premier-league-football-clubs-computer-analysts-managers-data-winning.

209 **Roberto Martinez, manager of Everton soccer club:** Ibid.

209 **When Picasso worked his "Bull" lithographs:** Details of bull given in: Lavin, Irving. "Picasso's Lithograph(s) 'The Bull(s)' and the History of Art in Reverse" *Art Without History*, 75th Annual Meeting, College Art Association of America, February 12–14, 1987.

210 **Einstein once said of scientific models:** Quoted by Sugihara, George. "On Early Warning Signs." *Seed Magazine*, May 2013. http://seedmagazine.com/content/article/on_early_warning_signs/.

211 **"The best material model of a cat":** Widely attributed to Wiener. Quote appears in: Rosenblueth, Arturo, and Norbert Wiener. "The Role of Models in Science." *Philosophy of Science* 12, no. 4 (1945): 316–321.

211 **In 1947, *Time* magazine published:** Chapin, R. M. "Communist Contagion." *Time*, April 1946. http://claver.gprep.org/fac/sjochs/communist-contagion-map.htm.

211 **a piece called "Europe from Moscow":** Chapin, R. M. "Europe from Moscow." *Time*, March 1952.

212 **When the pair bet together:** Borel, Émile. "A Propos d'Un Traite de Probabilities. Revue Philosophique." 1924. Quoted in Ellsberg, Daniel. *Risk, Ambiguity, and Decision* (New York: Garland Publishing, 2001).

213 **"How to Gamble If You Must":** Details of the course in: Bernstein, J. *Physicists on Wall Street and Other Essays on Science and Society* (New York: Springer, 2008).

213 **The MIT students therefore worked:** Details of the strategy given in: Mezrich, Ben. *Bringing Down the House: The Inside Story of Six MIT Students Who Took Vegas for Millions* (New York: Simon and Schuster, 2003).

214 **They later discovered the janitor:** Locker story given in: Ball, Janet. "How a Team of Students Beat the Casinos." BBC News Magazine, May 26, 2014. http://www.bbc.com/news/magazine-27519748.

214 **"I know very few people":** Author interview with Richard Munchkin, August 2013.

214 **In 2012, PhD student Will Ma:** Details and quotes from author interview with Will Ma, September 2014.

215 **Courses teaching the science of gambling:** The York University course was Bethune 1800: Mathematics of Gambling, taught in 2009–10, and the Emory course was MATH 190–000: Freshman Seminar: Math: Sports, Games & Gambling, taught in Fall 2012. Further details: http://garsia.math.yorku.ca/~zabrocki/bethune1800fw0910/ and http://college.emory.

216 **"That way of thinking about the world":** Author interview with Ruth Bolton, February 2013.

217 **"It is remarkable that a science":** Quoted widely, but originally given in: Laplace, P. S. *Théorie Analytique des Probabilitiés* (Paris: Courcier, 1812).

218 **"It wasn't as though streetwise Las Vegas gamblers":** Author interview with Bill Benter, July 2013.

INDEX

ADAM KUCHARSKI is a lecturer in mathematical modeling at the London School of Hygiene & Tropical Medicine and an award-winning science writer. Born in 1986, he studied at the University of Warwick before completing a PhD in mathematics at the University of Cambridge. He has published papers on topics ranging from statistics to social behavior, and has worked on the outbreak analysis of avian influenza and Ebola. Winner of the 2012 Wellcome Trust Science Writing Prize, he has contributed popular science articles to *Nautilus*, *BBC Focus*, and *Scientific American* magazines. He lives in London.